Springer Proceedings in Physics 429

Indexed by Scopus

The series Springer Proceedings in Physics, founded in 1984, is devoted to timely reports of state-of-the-art developments in physics and related sciences. Typically based on material presented at conferences, workshops and similar scientific meetings, volumes published in this series will constitute a comprehensive up to date source of reference on a field or subfield of relevance in contemporary physics. Proposals must include the following:

- Name, place and date of the scientific meeting
- A link to the committees (local organization, international advisors etc.)
- Scientific description of the meeting
- List of invited/plenary speakers
- An estimate of the planned proceedings book parameters (number of pages/articles, requested number of bulk copies, submission deadline).

Please contact:

For Americas and Europe: Dr. Zachary Evenson; zachary.evenson@springer.com
For Asia, Australia and New Zealand: Dr. Loyola DSilva; loyola.dsilva@springer.com

Sunao Kurimura · Tetsuya Yagi
Editors

Proceedings of Laser Display and Lighting Conference 2025

LDC 2025, June 17–20, Dublin, Ireland

Springer

Editors
Sunao Kurimura
National Institute for Materials Science
Tsukuba, Ibaraki, Japan

Tetsuya Yagi
Nichia Corporation
Anan, Tokushima, Japan

ISSN 0930-8989 ISSN 1867-4941 (electronic)
Springer Proceedings in Physics
ISBN 978-981-96-9781-6 ISBN 978-981-96-9782-3 (eBook)
https://doi.org/10.1007/978-981-96-9782-3

This Springer imprint is published by the registered company Springer Nature Singapore Pte Ltd.
The registered company address is: 152 Beach Road, #21-01/04 Gateway East, Singapore 189721, Singapore

Foreword

Welcome to the 14th Laser Display and Lighting Conference, LDC 2025 in Dublin!

LDC is an international conference on laser displays, laser lighting, and related technologies. The 1st LDC was held in Yokohama, Japan in 2012. After that, LDCs were held in Yokohama, Japan (2013, 2015, 2017–2019), in Taichung, Taiwan (2014), and in Jena, Germany (2016). The 9th–11th LDC were intended to be held in Yokohama, Japan, but switched to be on-line/hybrid on-line conferences (2020–2022) owing to the COVID-19 situation. The 12th and 13th LDC were held in Yokohama, Japan in-person (2023, 2024). Then, LDC 2025 is going to be held as an in-person conference in Trinity College, Dublin, Ireland from 17th to 20th June 2025. LDC 2025 is sponsored by the Optical Society of Japan, in cooperation with several academic societies and associations, and is operated by the Laser Display Research Group, the Optical Society of Japan.

LDC 2025 is intended to provide a central forum for the update and review of scientific and technical information on laser display and lighting covering a wide range of fields from fundamental research to systems and applications.

A total of 36 papers will be presented during the 4-day conference, consisting of two keynote talks, 25 invited papers, and nine contributed papers. A couple of post-deadline papers may be also accepted. All papers were peer-reviewed.

In 2024, teams from Barco NV and Nichia Corporation were recognized by the Academy of Motion Picture Arts and Sciences with a Technical Achievement Award (also known as a Technical Oscar ®) for their contributions to the development and integration of laser sources for digital cinema projectors. To mark this great advancement for laser display and lighting, the conference program committee has decided that the Keynote Session of LDC 2025 should include a commemorative presentation on 17th June.

Special sessions focusing on advanced XR and Metaverse technologies will be held on the 17th and 20th June, where we will have stimulating invited talks from expert speakers. An exciting special Joint session concerning 'Laser Technology for Moving Platforms' and "Light Sources and Components" will also be held with a number of distinguished speakers on 18th June. In these special sessions, state-of-the-art visible laser technology for XR, Metaverse and moving target applications, will be presented and discussed. After all the technical sessions, a ceremony for the LDC Best Paper Award and the LDC Student Award will be held for exceptional papers commended for their outstanding achievement.

We would like to extend our sincere thanks to all the presenters and participants of LDC 2025 for their contribution to the success of the conference. We also express our sincere thanks to LDC local committee members of Trinity College for their kind and warm-heart support. We hope that all the attendees enjoy the conference.

Prof. Kazuo Kuroda
Prof. Hiroshi Murata
Dr. Fergal Shevlin
Conference Co-chairs

LDC2025 Program

Tuesday, June 17, 2025

Opening Remarks

10:00 AM–10:30 AM

Keynote, LDC1

[LDC1-1 Keynote], 10:30 AM–11:15 AM
From Lab to Screen: the Journey of RGB Laser Projection and HDR in Cinema
Peter Janssens, Wouter D'Oosterlinck and Goran Stojmenovik
Barco

[LDC1-2 Keynote], 11:15 AM–12:00 PM
Blue and Green Laser Diodes for Cinema Projectors
Yoshitaka Nakatsu, Yoji Nagao, Tsuyoshi Hirao, Tomonori Morizumi, Kazuma Kozuru, Katsuhiro Kishimoto, Shintaro Harada, Teruyuki Morito, Tomoya Yanamoto and Shin-ichi Nagahama
Nichia Corporation

Light Sources and Components, LDC2-1-

[LDC2-1 Invited], 1:45 PM–2:15 PM
Recent progress on Gallium-Nitride-based Vertical-Cavity Surface-Emitting Lasers with curved mirror
Rintaro Koda[1], Tomohiro Makino[1], Kentaro Hayashi[1], Tatsushi Hamaguchi[2*], and Noriyuki Futagawa[1]

[1]Sony Semiconductor Solutions Corporations
[2*]Present affiliation: Kyusyu University

[LDC2-2 Invited], 2:15 PM–2:45 PM

Cyan, green, orange, and red structured light sources

Takashige Omatsu[1,2,3,4]
[1]Graduate School of Engineering, Chiba University
[2]Molecular Chirality Research Center, Chiba University
[3]Fujian Normal University
[4]National Yang Ming Chiao Tung University

[LDC2-3 Invited], 2:45 PM–3:15 PM

Mode-locked laser diode with highly-stacked InAs quantum dots

Kouichi Akahane[1], Satoshi Yanase[1,2], Atsushi Matsumoto[1], Toshimasa Umezawa[1], Naokatsu Yamamoto[1], Tomohiro Maeda[1,2], and Hideyuki Sotobayashi[2]
[1]National Institute of Information and Communications Technology
[2]Aoyama Gakuin University

[LDC2-4], 3:15 PM – 3:30 PM

InGaN-based Red Micro-Light-Emitting Diodes for Display Applications

Zhi Li[1], Brendan Roycroft[1], Changhao Li[1], Muhammet Genc[1], Abhinandan Hazarika[1], Bumjoon Kim[2], Soo Min Lee[2], Drew Hanser[2], Brian Corbett[1]
[1]Tyndall National Institute, University College Cork
[2]Veeco Instruments Inc.

XR (AR, MR, VR) and Metaverse LDC5-1-

[LDC5-1 Invited], 4:00 PM –4:30 PM

3D Displays Using Interleaved Fresnel Lenses

Hideki Kakeya and Hiroto Omori
University of Tsukuba

[LDC5-2 Invited], 4:30 PM – 5:00 PM

From Design to Display: Tackling Manufacturing Challenges in AR Waveguides

Juha Pietarinen and Joonas Pylväinenr
Optofidelity Oy

Wednesday, June 18, 2025

Light Sources and Components, LDC2-2-

[LDC2-5 Invited], 9:00 AM–9:30 AM

Polymer Optical Waveguide for Display Applications

Takaaki Ishigure
Keio University

[LDC2-6 Invited], 9:30 AM–10:00 AM

Development of Microlens Array Mold and Periodic Submicron Structure Mold by Using Single Crystalline Diamond Tools

Hiroshi Saito[1], Tsuneyuki Kobayashi[1], Yugo Sasaki[1], Satomu Kana-zawa[2], Takako Matsumura[2], Kazumi Sawamura[2], Eiji Shamoto[3] and Hirotsugu Yamamoto[4]
[1]Yamagata Research Institute of Technology
[2]IMUZAK Inc.
[3]Nagoya University
[4]Utsunomiya University

[LDC2-7], 10:00 AM–10:15 AM

Deformable Mirror Characterization for Efficient Fiber Coupling

Fergal Shevlin
Dyoptyka

[LDC2-8], 10:15 AM–10:30 AM

Holographic patterning of diffractive optical elements for beam shaping in lighting applications—a dual-hologram approach

Jorge Lasarte[1], Emma Branigan[1], Kevin Murphy[1], Izabela Naydenova[1], Jesús Atencia[2], Mª Victoria Collados[2] and Suzanne Martin[1]
[1]Technological University Dublin
[2]Engineering Research Institute of Aragon (I3A), Faculty of Science, Universidad de Zaragoza

Light Sources and Components, LDC2-3-

[LDC2-9, Invited], 11:00 AM–11:30 AM

MEMS-based Time, Wavelength, and Diffraction Multiplexed Image Transfer for Display Engine

Yuzuru Takashima, Chuan Luo, Gregory Nero, Yexin Pei, Tianyao Zhang, Xian-yue Deng, Ted Lee, and Parker Liu
Wyant College of Optical Sciences, University of Arizona

[LDC2-10 Invited], 11:30 AM–12:00 PM

Smart Contact Lenses

Dieter Cuypers and Herbert De Smet
Centre for Microsystems Technology, imec and Ghent University

[LDC2-11 Invited], 12:00 PM–12:30 PM

Visible near-to-eye display technology based on nonlinear frequency conversion

Peter G. R. Smith, Goronwy L. Tawy, Rex H. S. Bannerman, and Corin B. E. Gawith
Optoelectronics Research Centre, University of Southampton

Joint Session "Light Sources and Components" and "Laser Applications for Moving Platforms" LDC3-1-

[LDC3-1, Invited], 2:00 PM–2:30 PM
Measurement of properties of light generated in laser lighting
Ole Bjarlin Jensen[1], Jian Xu[2], Dennis Dan Corell[1] and Carsten Dam-Hansen[1]
[1]Diode Lasers and LED Systems Group, Department of Electrical and Photonics Engineering, Technical University of Denmark
[2]Lab of New Energy Materials and Devices, School of Physics and Electronic Information, Henan Polytechnic University

[LDC3-2, Invited], 2:30 PM–3:00 PM
Asynchronous Optical Sampling for High-Resolution LiDAR
Hollie Wright, Alexander J. M. Nelmes and Derryck T. Reid
Institute of Photonics and Quantum Sciences, School of Engineering and Physical Sciences, Heriot-Watt University

[LDC3-3, Invited], 3:00 PM–3:30 PM
Direct emitting diode laser modules below 630 nm for display applications
Katrin Paschke, Felix Mauerhoff, André Maaßdorf, David Feise, Oktay Senel and Gunnar Blume
Ferdinand-Braun-Institut (FBH)

Joint Session "Light Sources and Components" and "Laser Applications for Moving Platforms" LDC3-2-

[LDC3-4, Invited], 4:00 PM–4:30 PM
MEMS Technology as Enabler: From Accelerometer to LiDAR
Lj. Ristic[1], V. Milanovic[1], M.Vujosevic[2] and A. Kasturi[1]
[1]Mirrocle Technologies
[2]SensSpree

[LDC3-5, Invited], 4:30 PM–5:00 PM
Subwavelength Spatial Light Modulator powering the Next-Generation of Smart Eyewear
Andrzej Kaczorowski
Swave Photonics

Thursday June 19, 2025

Imaging and Lighting LDC 4-1-

[LDC4-1, Invited], 9:30 AM–10:00 AM
Volumetric Displays with Laser-Excitation: From Fingertip to Fist-Sized Graphics and Color
Kota Kumagai and Yoshio Hayasaki
Utsunomiya University
[LDC4-2, Invited], 10:00 AM–10:30 AM
Holographic Optical Elements for Application in Wound Healing Monitoring
Pamela Stoeva[1,2], Tatsiana Mikulchyk[1,2], Dervil Cody[1,2], Maria Antonietta Ferrara[3], Giuseppe Coppola[3], Suzanne Martin[1,2], and Izabela Naydenova[1,2]
[1]Centre for Industrial and Engineering Optics, School of Physics, Clinical and Optometric Sciences, Technological University Dublin
[2]Physical to Life Sciences Research Hub, Technological University Dublin
[3]Institute of Applied Sciences and Intelligent Systems (ISASI-NA)

Imaging and Lighting LDC 4-2-

[LDC4-3, Invited], 11:00 AM–11:30 AM
Pseudo-Direct LiDAR Based on an Ultra-High-Speed Charge-Domain Compressive Sensing Image Sensor
Keiichiro Kagawa
Shizuoka University
[LDC4-4, Invited], 11:30 AM–12:00 PM
Quantitative phase imaging under commercially available confocal microscopy
Naru Yoneda[1,2], Joe Sakamoto[3,4], Takumi Tomoi[5], Tomomi Nemoto[3,4,6], Yosuke Tamada[7,8,9], and Osamu Matoba[1,2]
[1]Graduate School of System Informatics, Department of System Science, Kobe University
[2]Center of Optical Scattering Image Science, Kobe University
[3]Biophotonics Research Group, Exploratory Research Center on Life and Living Systems
[4]Division of Biophotonics, National Institute for Physiological Sciences
[5]Faculty of Science and Technology, Department of Applied Biological Science, Tokyo University of Science
[6]School of Life Science, The Graduate University for Advanced Studies (SOKENDAI)
[7]Faculty of Engineering, Utsunomiya University
[8]Center for Optical Research and Education (CORE), Utsunomiya University

[9]Robotics, Engineering and Agriculture-Technology Laboratory (REAL), Utsunomiya University

[LDC4-5, Invited], 12:00 PM–12:30 PM

Application of Holography for Data Storage

Xiaodi Tan[1,2,3], Kazuo Kuroda[4], Jianying Hao[1], Shujun Zheng[1], Haiyang Song[1], Shenghui Ke[1], Yongkun Lin[1], Hongjie Liu[1], Ruixian Chen[1], Rupeng Yang[1], Dakui Lin[1,2], Xiao Lin[1],3 and Yuhong Ren[1]

[1]Information Photonics Research Center, College of Photonic & Electronic Engineering, Fujian Normal University

[2]Key Laboratory of Opto-Electronic Science and Technology for Medicine of Ministry of Education, Fujian Normal University

[3]Fujian Provincial Key Laboratory of Photonics Technology

[4]The University of Tokyo

Imaging and Lighting LDC 4-3-

[LDC4-6, Invited], 3:00 PM–3:30 PM

Dual-comb Spectroscopy Extended for Spatiotemporal Optical Sensing Applications using Structured Light with Orbital Angular Momentum

Akifumi Asahara and Kaoru Minoshima

Graduate School of Informatics and Engineering, The University of Electro-Communications

[LDC4-7], 3:30 PM–3:45 PM

Design of Meta-Surface Lens for THz-wave Object Imaging System in Security Application

Seiya Suzuki, Rizadi Sasmita Darwis and Hiroshi Murata

Graduate School of Engineering, Mie University

[LDC4-8], 3:45 PM–4:00 PM

2- and 3-Dimensional Imaging Using Field Interference Integration for Remote Sensing and Security Inspection Applications

Yamato Iwata, Riyogo Ishida, Arie Setiawan, Shu Sato, and Hiroshi Murata

Mie University

[LDC4-9, Invited], 4:00 PM–4:30 PM

Visible light generation by sum frequency conversion for quantum sensing and imaging

Sunao KURIMURA[1,2], Yuta KOCHI[2], Junko-ISHI HAYASE[2]

[1]National Institute for Materials Science

[2]Keio University

Friday June 20, 2025

XR (AR, MR, VR) and Metaverse LDC5-2-

[LDC5-3, Invited], 9:00 AM–9:30 AM
Projection Mapping Technologies for Seamless Spatial Augmentation
Daisuke Iwai
The University of Osaka

[LDC5-4, Invited], 9:30 AM– 10:00 A M
Enhancing Visual Realism with Projector-Camera Feedback and Light-Field Projection
Toshiyuki Amano
Wakayama University

[LDC5-5, Invited], 10:00 AM–10:30 AM
Contact Lens Display Using Holographic Image Generation
Yasuhiro Takaki
Tokyo University of Agriculture and Technology

Novel and Emerging Technologies LDC6

[LDC6-1], 11:00 AM–11:15 AM
Temperature distribution of AMOLED panel with the condiseration of IR-drop
Qibin Feng[1-2], Baoceng Wang[1-2], Zi Wang[1-2], Guoqiang Lyue[1-2]
[1]Hefei University of Technology
[2]Anhui Province Key Laboratory of Measuring Theory and Precision Instrument, School of Instrument Science and Optoelectronics Engineering, Hefei University of Technology

[LDC6-2], 11:15 AM–11:30 AM
Multi-Gbps Visible Light Communication with High-Efficiency, Low-speckle Contrast White Laser Light
Joseph P. Lazzaro[1], Xiaoxin Wang[1], Charles J. Carver[2], Nicholas Shade[1], Hadleigh Schwartz[2], Yanchen Liu[2], Xia Zhou[2], Jifeng Liu[1], Eric R. Fossum[1]
[1] Dartmouth College
[2] Columbia University

[LDC6-3], 11:30 AM–11:45 AM
Evaluation of Module for Alignment-Free Diffuser Film Composed of Dielectric Multilayers and Microprisms for Mini LED Backlight
Marii Nishikawa[1,2], Yukio Taniguchi[1], Masahiro Goto[1], Yasuyuki Ohyagi[1], Yoshihiro Kanai[1], Masayuki Sekido[1], Hirotsugu Yamamoto[2]
[1]Dai Nippon Printing Co., Ltd.
[2]Utsunomiya University

[LDC6-4], 11:45 AM–12:00 PM
Elimination of Hot Pixels in Measuring the Point Spread Function of Aerial Display
Takumi Watanabe, Kazuaki Takiyama, Shiro Suyama, and Hirotsugu Yamamoto
Utsunomiya University

Closing Remarks

12:00 PM–12:30 PM

Contents

Part V XR (AR, MR, VR) and Metaverse LDC5

Part VI Novel and Emerging Technologies LDC6

Part I
LDC Keynote LDC1

From Lab to Screen: the Journey of RGB Laser Projection and HDR in Cinema

Peter Janssens[1[0009-0007-6778-6502]], Wouter D'Oosterlinck[1[0009-0003-7613-7273]] and Goran Stojmenovik[1[0009-0009-7563-3770]]

[1] Barco – Entertainment Division, Beneluxpark 21, 8500 Kortrijk, Belgium
peter.janssens@barco.com

Abstract. Barco began researching laser technology for cinema projectors in 2007, aiming to enhance image quality and lifetime. Initial challenges included speckle, power scaling, and efficient 3D solutions. By 2014, Barco introduced the first integrated RGB laser cinema projector.

Initially, the adoption of RGB laser projectors in the market was hindered by high costs and stringent laser safety standards, restricting their use primarily to premium cinemas. However, adaptions of safety regulations, advancements in laser diodes and subsequent cost reductions have led to the widespread dominance of RGB laser and laser phosphor projectors in today's market.

Until recently, laser projectors couldn't achieve the contrast ratio and brightness level needed for HDR images on large screens. This changed with Barco's HDR light steering technology introduced in 2024.

Keywords: Laser Projection, Cinema, Light Steering.

1 Introduction

The cinema market, a subset of the projection market, adheres to specific security and image quality standards set by the Digital Cinema Initiatives (DCI), a consortium of major motion picture studios. These standards are essential for displaying movies from DCI members and have become the industry norm.

Initially, cinema projectors were analog, using film for content delivery. The shift to digital projection began around 2000 and gained momentum with the release of 3D movies like Avatar (2009).

Xenon lamps were the primary light source for cinema projectors. Powerful Xenon lamps needed replacement every 1,000 hours due to reduced output and flicker. Lasers, with lifetimes up to 30,000 hours, emerged as the ideal light source, enhancing reliability, brightness, and image quality (better color gamut, uniformity, and contrast). The transition to lasers would significantly impact the cinema market.

S. Kurimura and T. Yagi (eds.), *Proceedings of Laser Display and Lighting Conference 2025*, Springer Proceedings in Physics 429,
https://doi.org/10.1007/978-981-96-9782-3_1

2 Development of the First RGB Laser Cinema Projector

Barco began laser projection research in 2007, developing four prototypes over five years. Early tests with a low-power, breadboard-mounted projector highlighted three key challenges: reducing speckle, achieving efficient power scaling for high brightness, and ensuring high-quality 3D projection.

2.1 Speckle Reduction

Speckle refers to the granular interference patterns in laser-projected images. The origins and reduction methods of speckle are well-documented in literature, e.g. [1]. Mitigation techniques involve displaying uncorrelated speckle patterns by introducing random phase differences in the laser beam (e.g., multiple emitters or moving diffusers - known as angle diversity), using laser sources with a wide spectrum, or depolarizing the projector's emitted light.

In 2007, no reliable method to measure speckle or information on acceptable speckle levels existed. Barco developed a speckle contrast measurement procedure for its own research that closely mimicked human vision and published initial results on speckle perception levels [2]. This method formed the basis for LIPA's recommended practice for laboratory speckle measurements (2015) [3], standardizing the process for the industry. However, data on human speckle perception remains limited, so Barco relied on self-organized tests in cinematic environments to define target speckle levels for its products.

Angle and polarization diversity are design parameters of RGB laser projectors, but both methods hit a maximum speckle reduction limit determined by the F-number of the projector and spectral broadening is needed for optimal results. Initially, obtaining lasers with different wavelengths was challenging, so speckle pattern decorrelation experiments were performed by temperature detuning of near-IR laser diodes to determine the minimum wavelength difference needed for speckle reduction on various screens. The required spectral broadening varies by color and e.g. for green lasers, a 1 nm separation between individual lasers was a good compromise between technical achievability and image quality for common white projection screens. Only green lasers based on frequency doubling of IR diodes met this specification. For blue and red, the existing direct laser diodes allowed sufficient spectral broadening.

2.2 High Brightness and System Efficiency

Lasers allow for fiber delivery of light into existing projection systems. This approach places the lasers in a separate rack, simplifying cooling, driving, and making the system more modular. It also decouples the opto-mechanical tolerances of the lasers from the projector's optical system via flexible optical fibers.

However, fiber coupling results in significant light losses and raises reliability and safety concerns if a fiber breaks. While some of Barco's initial demonstrators used fiber coupling for convenience, free-space coupling was deemed the best approach for RGB laser cinema projectors to achieve the desired brightness at the lowest cost. Designing a free-space laser source of 60,000 lm (60 klm) required extensive Monte Carlo tolerance modeling for the combining optics and a dedicated cooling and driving solution for 600W of laser light distributed across hundreds of individual lasers.

2.3 3D Stereo Projection Using Six Primaries

With lamp projectors, 3D stereo projection methods based on polarization or color filters suffer from significant efficiency losses, resulting in dim 3D compared to 2D images. The color filter method e.g. used two non-overlapping sub-spectra of the lamp to project images for the left and right eyes. To minimize stereo cross-talk, sufficient spacing between the spectra is required, leading to each eye receiving only about 12% of the projector's brightness in 2D. Typical spectra for the left and right eye on a Xenon lamp projector are shown in the left panel of **Fig. 1**.

Using six primaries, or two sets of RGB lasers, significantly reduces light losses in color filter 3D images. There is no light falling between filters, and while the left eye image is displayed, the right eye lasers can be switched off. This makes the projector 4 times more energy-efficient in 3D mode compared to a lamp projector. This method is often referred to as 6P 3D.

In collaboration with selected laser suppliers, Barco defined two sets of RGB wavelengths to enhance image quality (e.g., cross-talk and color uniformity) of 6P 3D compared to lamp-based solutions. Other considerations included required spectral width for speckle reduction, luminous efficacy of the primaries, white balancing, cost, and technical limitations of available laser wavelengths. The resulting spectra are shown in **Fig. 1**.

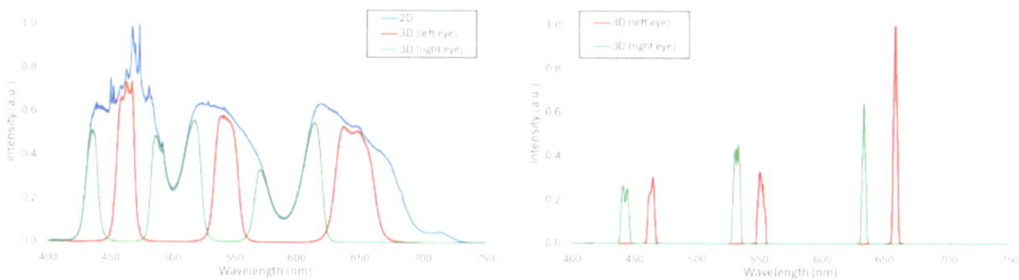

Fig. 1. Spectra for color filter based 3D compared to 2D in case of a Xenon lamp projector (left) and left and right eye spectra for a 6P 3D capable projector (right).

2.4 The First RGB Laser Cinema Projector: DP4K-60L

Years of research in laser technology, despeckling, free-space combination, and 6P 3D culminated in Barco's DP4K-L projectors, launched in 2014. At the time, they were the brightest integrated RGB laser projectors available, offering 6P 3D from a single projection head and delivering speckle performance suitable for demanding cinema applications.

The DP4K-L series includes four projectors ranging from 20 to 60 klm. The imaging component of the DP4K-L is similar to its lamp-based counterparts (DP4K-B series), utilizing a standard 3-DLP engine with 4K DMDs (4096x2160 pixels).

The lasers are grouped by color into clusters containing both left and right eye lasers. Mirrors and lenses bundle the output of each cluster into a single polarized beam. **Fig. 2** shows the blue cluster. The optical output of the clusters is similar in beam size and divergence across all three colors. Each cluster also includes laser drivers and cooling blocks. An active compressor-based system provides cold cooling liquid to the laser clusters and DMDs to ensure lifetimes exceeding 30,000 hrs.

In the highest configuration, the laser light source consists of four clusters per color, with fewer clusters in lower brightness variants. Two clusters of the same color are first combined based on polarization, then the three colors are combined into a single white beam. Finally, two such white beams enter the uniformization optics side by side. The entire optical path of the projector is enclosed by panels with safety interlocks and can remain in place during alignment making the projector compliant with the laser safety standard IEC 60825.

Fig. 2. The DP4K-60L (left) and its blue laser cluster (right)

3 Mainstream RGB Laser Projection: SP4K-series

3.1 Evolution of Lasers

Initially, RGB laser projectors were used in premium cinemas due to their cost and complexity. This would change with the introduction in 2016 of Watt-class green laser diodes, of which first lasing was achieved in 2012 [4]. Before this, green lasers were inefficient and complex to drive due to the second harmonics generation process requiring individual laser calibration. Green laser diodes, on the other hand, use conceptually the same laser drivers as blue and red diodes, simplifying the driving of the overall light source. In addition, blue and green laser diodes have made important gains in efficiency, which allowed for radiator-based cooling systems instead of active cooling with chilled water. Red lasers, however, still require active cooling for optimal performance, which can be achieved by a TEC cooling element as interface between the laser and the water-cooling system.

The laser market saw standardization with multiple suppliers using 8 TO9-cans on a common substrate, known as a laser bank, making the optomechanical system uniform across colors, which simplified the light source design. Additionally, laser diodes became available in standardized multi-emitter packages with 20 or more lasers per housing. Further consolidation occurred, allowing all colors to be sourced in the same form factor from a single supplier.

3.2 Laser Illuminated Projector Association (LIPA)

An important contribution to the mainstream adoption of laser-based projectors was achieved by LIPA (Laser Illuminated Projector Association), an industry organization of the world's industry leaders in projection technology formed in 2011 acting as a single voice in the laser illuminated projector industry. Its main purposes are promoting and reducing barriers to laser projector adoption, striving to rationalize and harmonize various laser regulations worldwide, providing a forum to develop evaluation methods and educate key constituencies on laser projector technology.

Laser projectors are not more dangerous than lamp projectors having the same radiance. However, initially all laser projectors needed to comply to the laser safety standard IEC 60825, which is much stricter than the IEC 62471 for lamps. By the work of LIPA, first the laser standard was adopted to allow exclusion for laser illuminated projectors, and since 2015 nearly all laser projectors are regulated under the updated IEC 62471-5 standard, relaxing the design and installation regulations.

3.3 Modular RGB Laser Light Sources

The availability of suitable lasers and resolution of regulatory issues enabled the introduction of the Barco SP4K-series RGB laser projectors with 4K resolution to the mainstream cinema market in 2019. A modular design was adopted, allowing one light source concept to support 9 projector models ranging from 12 to 55 klm. The design prioritized cost and serviceability while ensuring a long lifetime for worry-free operation.

The light source is divided into small, serviceable units called laser plates (see **Fig. 3**), each containing 2 or 3 laser packages with 8 diodes each. These are mounted on an aluminum plate that serves as both a mechanical interface to the projector's optomechanics and a cooling interface to the water-cooling circuit. An integrated PCB serves as an interconnection for laser driving and it has memory on-board for identification and runtime logging. All colors share the same laser plate design.

Power scaling is achieved by combining laser plates using spatial, polarization, and color filter methods. Each step in power scaling has its own building block. The minimum configuration uses one laser plate with 2 laser packages per color, while the maximum uses 24 laser packages on 8 laser plates.

The modular approach offers several advantages. Exhibitors don't need to stock many expensive spare light sources for different projector power levels; 8 different laser plates can maintain all RGB projectors, regardless of output power. The standardization simplifies the automated production of base components and the logistics, as the final light source assembly can occur later in the production process.

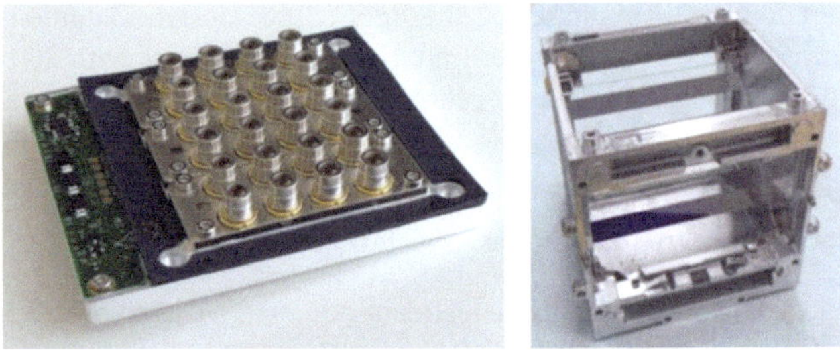

Fig. 3. The laser plate, and a spatial power scaling block of the SP4K series.

4 High Dynamic Range

4.1 Limitations of Cinema Projectors for HDR

Typical intraframe contrast ratios of cinema projectors are around 2,000:1. Higher contrasts up to 5,000:1 can be achieved with higher F-number optical systems, like Barco's SP4K-27HC, though this reduces system transmission efficiency. Sequential on/off contrast ratios can be enhanced with content-based dynamic dimming, revealing more details in overall dark scenes only. Other approaches include a prismless optical architecture that achieved a 10,000:1 contrast ratio [5], or a system using a cascade of two sets of DMDs, increasing intraframe contrast ratios to over 1M:1 [6].

By design, all concepts mentioned above have a (first) DMD that is uniformly illuminated and that sends light to a dump in the dark regions of the frame. In this way, displaying High Dynamic Range (HDR) content up to 300 nit on large cinema screens is costly and poses thermal management challenges for the DMDs, because of the required amount of laser power.

4.2 Principle of Light Steering

Barco's HDR light steering technology uses a different approach. While the image is still formed by a 3-DLP optical engine, phase modulators upstream in the light path direct light from darker to brighter regions of the image. This allows for higher local brightness levels. Additionally, steering light away from darker areas lowers the black level, revealing more details in the shadows. This results in more efficient use of laser light, and it significantly extends the projector's dynamic range.

For HDR light steering the image content is first converted into an illumination target distribution for the DMD. Then, a phase pattern is calculated such that the phase modulator generates this distribution by diffraction (see **Fig. 4**). Finally, the content is adapted towards the illumination profile of the DMD, such that the correct HDR image frame is shown on the screen.

Fig. 4. From left to right: the HDR original image frame, the global intensity profile on the DMD and associated phase pattern on the phase modulator.

In addition to the different content flow, light steering technology requires a dedicated powerful and low étendue light source to illuminate the phase modulator, ensuring sufficient beam steering power for high brightness levels. The phase modulator's maximum steering angle is determined by its pixel pitch, and only the zone between diffraction orders -1 and 1 is used for steering. Considering beam propagation factors, the light source's étendue must be over 1,000 times smaller than that of the DMD that forms the final image.

4.3 First Light Steering Cinema Projector: LS4K

The Barco LS4K, the first projector with light steering, has been rolled out since late 2024. It supports HDR playback on screens up to 18m wide. The projector features a highly collimated laser source, phase modulators, and dedicated optics to achieve the peak brightness levels, needed for HDR using the principle of light steering, as well as a uniform baseline illumination to obtain the standard full screen white level for SDR operation.

Light steering requires a specific light source design to achieve high output power within the low étendue available for the phase modulator. First, a lens collimates and aligns the pointing of the individual laser beams with sub-mrad accuracy. Then, 10 laser beams are combined using knife edging, forming a 10 by 1 laser module. These modules are further stacked to reach the desired power level. For red, the most demanding color, the final beam consists of 120 highly collimated and closely spaced laser beams.

The LS4K is optimized to show both SDR and HDR movies in a cost-effective way. For SDR content, light steering is disabled and the DMD is illuminated uniformly, making the projector function like a standard one. In HDR mode, light steering is used to boost the brightness in the necessary screen areas.

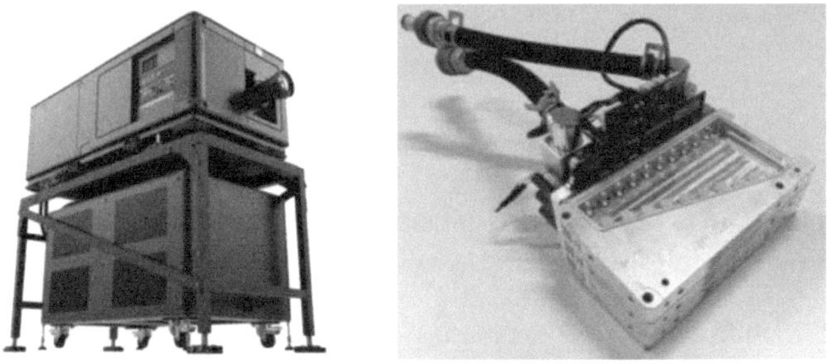

Fig. 5. Left: LS4K, Barco's first HDR projector based on light steering (left); a stack of three 10 by 1 laser modules (right).

5 Summary and Outlook

RGB laser cinema projectors have significantly impacted the cinema industry. Barco began investigating laser technology in 2007, leading to the world's first and brightest integrated RGB projector in 2014. Initially, these projectors provided reliability and high-quality images mainly in premium theaters. However, advancements in laser diode technology and their integration into modular projectors have enabled laser adoption in mainstream cinemas since 2019. Innovation continues, with HDR expected to further enhance the movie-going experience in the coming years.

Acknowledgments. Barco's laser research received support from the Flemish government (VLAIO).

References

1. Goodman, J.: Speckle phenomena in optics: theory and applications. Roberts and Company, Englewood (2007).
2. Roelandt, S., Meuret, Y., Craggs, G., Verschaffelt, G., Janssens, P., & Thienpont, H.: Standardized speckle measurement method matched to human speckle perception in laser projection systems. Opt. Express 20, 8770-8783 (2012).
3. LIPA Speckle Metrology Workgroup: Recommended Practice: Measurement of Speckle in Laboratory Conditions. LIPA Homepage, https://www.lipainfo.org/laser-illumination-sources/speckle-metrology, last accessed 2025/01/02.
4. Nakatsu, Y., Nagao, Y., Hirao, T., Hara, Y., Masui, S., Yanamoto, T., Nagahama, S.: Blue and green InGaN semiconductor lasers as light sources for displays. In: Proc. SPIE 11280, Gallium Nitride Materials and Devices XV, pp. 1-7 (2020).
5. Perkins, M., Dewald, D., Wainwright, N., Two-stage light modulation for high dynamic range, US 9,232,172 B2 patent (2013).
6. Silverstein, B., Kurtz, A., Bietry, J., Nothard, G.: A Laser-Based Digital Cinema Projector. In: SID Symposium Digest of Technical Papers, 42, pp. 326-329 (2011).

Blue and Green Laser Diodes for Cinema Projectors

Yoshitaka Nakatsu, Yoji Nagao, Tsuyoshi Hirao, Tomonori Morizumi,

Kazuma Kozuru, Katsuhiro Kishimoto, Shintaro Harada, Teruyuki Morito,

Tomoya Yanamoto and Shin-ichi Nagahama

Nichia corporation, 491 Oka, Kaminaka-cho, Anan, Tokushima 774-8601, Japan
yoshitaka.nakatsu@nichia.co.jp

Abstract. This paper reports the latest device performance of high-power blue and green edge-emitting laser diodes (LDs) for cinema projectors. The Gallium Nitrides based epitaxial layers of LDs were grown by metal organic chemical vapor deposition (MOCVD). Fabricated every LD chip was mounted on a heat sink using a junction down method in a TO-Φ9 mm package. We optimized the epitaxial and the device structures for high efficiency and high optical output power. A new developed 465 nm blue LD showed the optical output power, the voltage and the wall-plug efficiency (WPE) of 6.57 W, 3.86 V and 48.7 % at the forward current of 3.5 A under continuous wave (CW) operation. The peak WPE of the 465 nm LD was 50.0 % at the forward current of 2.3 A. And our developed 525 nm green LD showed the optical output power, the voltage and the WPE of 1.97 W, 4.12 V and 25.2 % at the forward current of 1.9 A under CW operation. The peak WPE of the 525 nm LD was 26.4 % at the forward current of 1.2 A.

Keywords: InGaN, Laser Diodes, Blue Laser Diodes, Green Laser Diodes.

1 Introduction

Gallium nitride has been attracting attention as a material for electronic devices, light-emitting diodes (LEDs), and semiconductor lasers (LDs), etc. Since we realized the world's first room-temperature continuous wave (CW) emission of a gallium nitride-based semiconductor laser diode in 1995 [1], various applications have been attempted. The adoption of LDs for projector applications has been progressing rapidly over the past 10 years. We developed high-efficiency blue and green LDs for laser display light sources, with the WPE of 21.2% for 525 nm green LDs and 48.1% for 455 nm blue LDs [2]. In addition to nitride-based LDs, efficiency improvements of red LDs have been progressing, with WPE up to 43% of 638nm AlGaInP based LDs by other groups [3]. Thus, highly efficient RGB LDs have been realized. Because visible light LDs have the advantage of narrow spectral width, light sources of RGB LDs expand the color gamut of cinema and makes the color reproduction closer to that of human eyes. In recent years, LDs have been adopted as light sources for cinema projectors.

S. Kurimura and T. Yagi (eds.), *Proceedings of Laser Display and Lighting Conference 2025*, Springer Proceedings in Physics 429,
https://doi.org/10.1007/978-981-96-9782-3_2

2 Device Structure

Schematic structures of our blue and green LD chips (see Fig. 1). The blue LD has epitaxial layers including n-type, active and p-type layers, which were grown by metalorganic chemical vapor deposition (MOCVD) on a C-plane free-standing GaN substrate. These layers are so-called a separate confinement heterostructure based on cladding, guiding and quantum well layers. And a ridge was formed by etching the p-type layers to the width of 45 μm for a 465 nm blue LD. In the case of a 525 nm green LD, the ridge width was formed to 20 μm. Electrodes of the n-type and p-type were formed at the substrate and the ridge respectively. And the front and rear sides were obtained by cleavage at the m-plane surface. These faces were covered by dielectric mirrors. Considering optical gain and mirror loss, the cavity length was set at 1.5 mm. Finally, the LD chip was mounted on a heat sink using a junction down method in a TO-Φ9 mm package. The characteristic evaluation has been performed with the LD structure fabricated in this way, and the results of this characteristic evaluation are described in following sections.

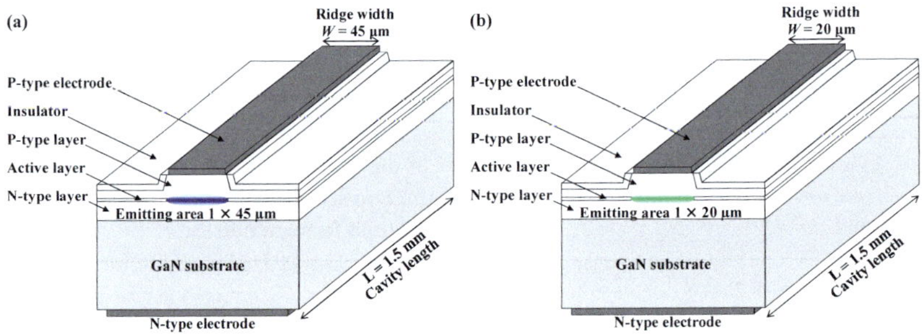

Fig. 1. Schematic structures of (a) 465 nm blue and (b) 525 nm green laser diode chips

3 LDs for Cinema Projectors

3.1 Blue LDs for Cinema Projectors

The first 460 nm blue LD had the optical output power of 5 mW in 2001 [4]. Recently we developed 455 nm blue LD with the optical output power of 5.99 W and the WPE of 52.4 % for projectors [2]. We have been developing 465 nm LDs for cinema projectors. The latest results are shown below.

3.2 465 nm blue LD.

We fabricated a new high-power blue LD. The LD has the peak wavelength of approximately 465 nm at the forward current of 3.5 A (see fig. 2).

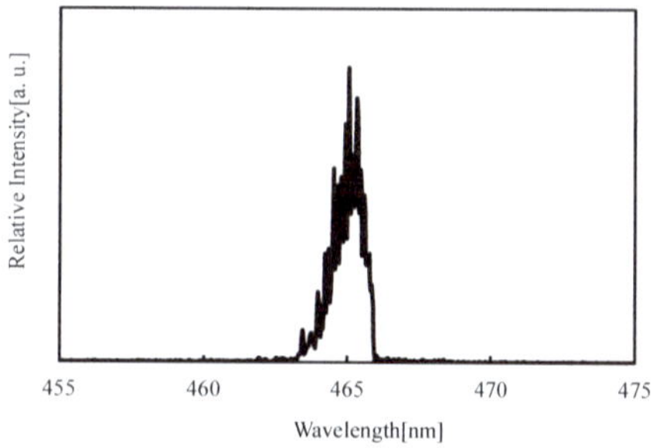

Fig. 2. Emission spectrum of the 465nm blue LD under 3.5 A CW operation at 25 °C

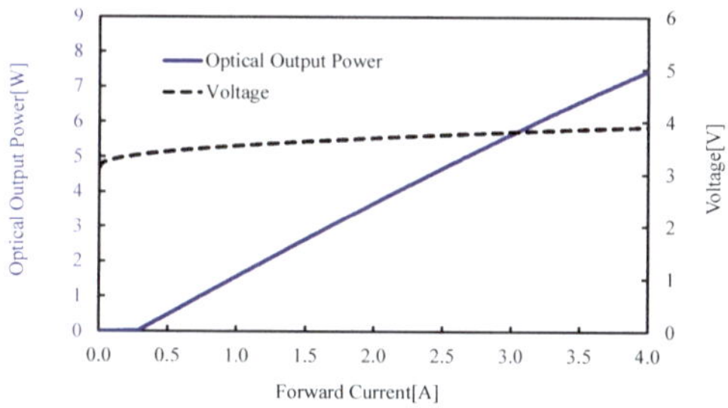

Fig. 3. L-I and V-I characteristics of the 465 nm blue LD under CW operation at 25 °C

The output power-current (L-I) and voltage-current (V-I) characteristics of the blue LD during CW operation at the case temperature of 25°C are shown (see Fig.3). The threshold current of the blue LD is 0.28 A. The blue LD shows the optical output power and the voltage of 6.57 W and 3.86 V at the forward current of 3.5 A under CW operation.

3.3 525 nm Green LDs

Subsequent paragraphs, however, are indented. For use in digital cinema projectors, watt-class green LDs were needed to satisfy the chromaticity of digital cinema. In 2012, we reported the first watt-class 525 nm LDs, and the optical output power and the wall plug efficiency of 525 nm LDs were 1W and 14.1%. We have already reported further efficiency development of 525 nm green LDs, which had the wall plug efficiency of 21.2% in 2020 [5]. These epitaxial layer structures of LDs were directly grown on free-standing C-plane GaN substrates. In our recent report, the WPE of green LD at 525 nm achieves 25.2% [2].

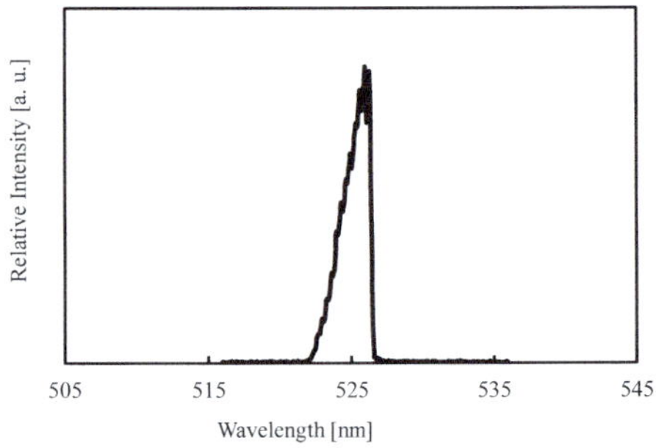

Fig. 4. Emission spectrum of the 525nm green LD under 1.9 A CW operation at 25 °C

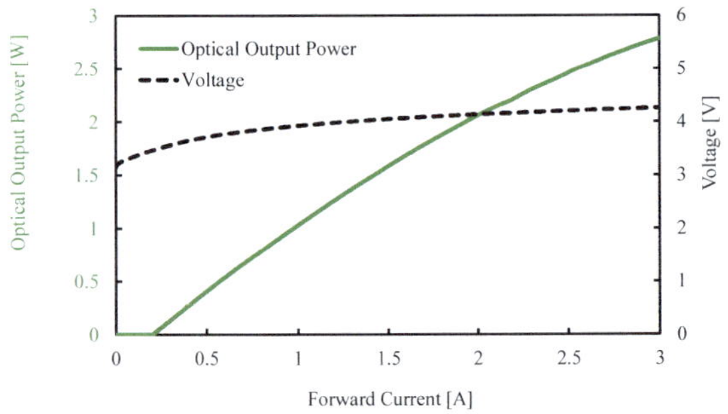

Fig. 5. I-L and I-V characteristics of the 525nm green LD under CW operation at 25 °C

We fabricated a new high-power green LD by forming a single ridge, and the ridge width was 20 μm. The peak wavelength of the green LD is about 525 nm at the forward current of 1.9 A under CW operation (see Fig. 4), and the threshold current of this LD is 0.20 A (see Fig. 5). The 525 nm LD shows the optical output power and the voltage of 1.97 W and 4.12 V at the forward current of 1.9 A under CW operation. The WPE reaches 25.2 % at the current of 1.9 A. And the peak WPE of the 525 nm LD is 26.4 % at the forward current of 1.2 A.

3.4 532 nm Green LDs

532 nm Green LDs are increasingly being used as pure green light sources for cinema applications. Our developed 532 nm green LD showed the optical output power and voltage of 1.64 W and 4.26 V at the forward current of 1.9 A under CW operation [5]. The WPE of the 532 nm LD was 20.2 % at 1.9 A.

4 Conclusion

We have shown the characteristics of our latest blue LDs grown on c-plane free-standing GaN substrates by MOCVD. By optimizing the epitaxial layers, device structures, and packaging, we successfully developed high-efficiency and high-power LDs. The optical output, voltage, and WPE of the 465 nm blue LD for cinema projectors were 6.57 W, 3.86 V, and 48.7 % at a 3.5 A forward current under CW operation. High-power 525 nm and 532 nm green LDs were developed. 525 nm LDs show optical output, voltage, and wall plug efficiency of 1.97 W, 4.12 V, and 25.2 % at a forward current of 1.9 A in CW operation. The WPE and optical output of the high-power LDs were constantly improved, covering a wide visible light range. These results contribute to the cinema projector utilization of laser diodes. We will continue to contribute to the improvement of the efficiency of visible light LDs and cinema projector applications.

References

1. Nakamura, S., Senoh, M., Nagahama, S., Iwasa, N., Yamada, T., Matsushita, T., Kiyoku, H., Sugimoto, Y.: InGaN-Based multi-quantum-well-structure laser diodes. Jpn. J. Appl. Phys. 35, L74-L76 (1996)
2. Kishimoto, K., Hirao, T., Morizumi, T., Nagao, Y., Nakatsu, Y., Yanamoto, T., Nagahama, S.: Development of highly efficient blue and green edge-emitting laser diodes. In: Proceedings of SPIE Photonics West 2024, vol.12886, pp. 1288609 1-6 (2024)
3. Kawanaak, S., Kitamura, S., Miyamoto, S., Hashizume, M., Yano, K., Hagimoto, M.: High power red laser diodes for display applications. In: SID International Symposium Digest of Technical Papers, vol. 53, no. 1, pp. 953–955 (2022).
4. Nagahama, S., Yanamoto, T., Sano, M., Mukai, T.: Characteristics of InGaN laser diodes in the pure blue region. Appl. Phys. Lett. 79(13), 1948–1950 (2001).
5. Nakatsu, Y., Nagao, Y., Hirao, T., Hara, Y., Masui, S., Yanamoto, T., Nagahama, S.: Blue and green InGaN semiconductor lasers as light sources for displays. In: Proceedings of SPIE, vol. 11280, pp. 112800S 1-7 (2020).

Part II
Light Sources and Components LDC2

Recent progress on Gallium-Nitride-based Vertical-Cavity Surface-Emitting Lasers with curved mirror

Rintaro Koda[1], Tomohiro Makino[1], Kentaro Hayashi[1], Tatsushi Hamaguchi[2*], and Noriyuki Futagawa[1]

[1] Sony Semiconductor Solutions Corporations, Atsugi 243-0014, Japan
[2*] Present affiliation; Mie university, Japan
rintaro.koda@sony.com

Abstract.
Highly efficient GaN-VCSELs emitting in blue wavelength were obtained. The peak wall plug efficiency was 22.2 %. This talk also reviews the latest progress of two-dimensional blue VCSEL array producing watt-class output power.

Keywords: GaN-VCSEL, blue VCSEL array.

In recent years, Vertical-Cavity Surface-Emitting Lasers (VCSELs) have seen expanded applications in sensing and optical communications, e. g., in smartphones and data centers. These VCSELs emit near infrared wavelength. There would be a broad range of potential applications including displays, lighting, material processing, bio-sensing, and visible light communication, when VCSELs emitting visible wavelength are realized. GaN material system can provide optical gain in the visible spectrum range and performance of GaN-based VCSELs, after 1st demonstration of CW lasing operation at room temperature in 2008[1], progressed significantly.

There are two main approaches to realize GaN-based VCSEL. The first approach involves the use of dielectric Distributed Bragg Reflectors (DBRs) and the second approach utilizes epitaxial growth of semiconductor DBRs as bottom side mirror. Both types of structure have demonstrated over 10 mW output power and 10 % wall plug efficiency [2-5]. We have been developing a unique VCSEL where the backside of GaN-substrate has curved mirror [6]. In this talk, we will report the previous and the latest progress of our research.

We have previously reported various results for blue VCSELs, including a low threshold current of 0.25 mA [6], a narrow emission angle of 3.9 degrees [7] achieved through precise control of the mirror curvature. Additionally, we demonstrated single longitudinal mode operation with an SMSR of 40dB by modulating the mirror reflectivity [8]. One of the research focuses of the authors is on green VCSELs, where we have previously reported continuous wave operation of green VCSELs grown on the {20-21} semi-polar plane, achieving a high efficiency of 3.6% [5,7] and white light generation solely with VCSELs.

S. Kurimura and T. Yagi (eds.), *Proceedings of Laser Display and Lighting Conference 2025*, Springer Proceedings in Physics 429,
https://doi.org/10.1007/978-981-96-9782-3_3

Our recent progress of efficiency improvement by several fabricating processes and the respective laser thicknesses were optimized for blue VCSEL. The maximum wall plug efficiency was 22.2 %. We also fabricated two-dimensional blue VCSEL array and demonstrated watt-class high power operation.

Acknowledgments: A part of this work is based on results obtained from a project, JPNP21005, subsidized by the New Energy and Industrial Technology Development Organization (NEDO).

References

1. Y. Higuchi, et. al., Room-Temperature CW Lasing of a GaN-Based Vertical-Cavity Surface-Emitting Laser by Current Injection: Appl. Phys. Express, 1, 121102 (2008).
2. H. Nakajima., Appl. Phys. Express, 12, 084003 (2019).
3. M. Kuramoto, et. al., Appl. Phys. Express, 13, 082005 (2020).
4. R. Watanabe, et. al., Appl. Phys. Lett. 124, 131107 (2024).
5. Ito. M., In Proceedings of the 29th International Display Workshops, Fukuoka, Japan, 14 December (2022).
6. Hamaguchi, T., et. al., Sci. Rep. 8, 10350 (2018).
7. Hayashi, K., IEEE Photonics J., 14, 1536905 (2022).
8. Kearns. A. J., Appl. Phys. Express, 15, 072009 (2022).
9. Hamaguchi, T., Appl. Phys. Express, 13, 041002 (2020).

Cyan, green, orange, and red structured light sources

Takashige Omatsu[1,2,3,4]

[1] Graduate School of Engineering, Chiba University, Japan
[2] Molecular Chirality Research Center, Chiba University, Japan
[3] Fujian Normal University, China
[4] National Yang Ming Chiao Tung University, Taiwan

takashige.omatsu@gmail.com
omatsu@faculty.chiba-u.jp

Abstract. We herein report on the multicolor (cyan, green, orange, red, and deep-red) visible structured light sources based on compact and robust diode-pumped Pr3+ doped lasers. These structured laser source allow the development of ultra-fine 2-dimensional or 3-dimensional microfabrication, ultra-high resolution laser scanning microscopy, and advanced light-sheet imaging systems.

Keywords: Structured light, Pr^{3+} doped lasers, Singular optics, Topology

Laser and laser optics research has always spent the efforts to produce only a fundamental Gaussian mode with high efficiency and high quality. Beyond this conventional approach, "structured light" with the engineered wavefront or polarization in 2-dimnsional (2D) or 3-dimensional (3D) real space has been rapidly growing. An optical vortex, the most used structured light, carries an annular spatial form and an orbital angular momentum (OAM) associated with its helical wavefront.

At present, the structured light further encompasses particle-like optical fields, so-called optical quasiparticles, possessing the topologically protected polarization (Stokes vectors) textures (space-invariant polarization textures even under turbulent media).

Such structured light has been potentially applied in a myriad of applications, such as free-space optical/quantum communications under turbulent conditions, spatial multiplexing optical recording, super resolution fluorescence microscopy, and polarization nanoscale-metrology through exotic light-matter interactions.

A Pr^{3+} doped laser materials with strong absorption in the blue region (~442 nm) exhibit the emission peaks (cyan (~490 nm), green (~520 nm), orange (~610 nm), red (~640 nm), dark-red (~720 nm)) in the visible region. To date, several groups demonstrated the direct generation of red optical vortex modes from a Pr^{3+}:YLF laser by employing various approaches, such as an annular pumping and off-axis pumping configurations [1,2].

Furthermore, the direct generation of green, orange and red Bessel beams with OAM has been demonstrated in a diode-pumped Pr^{3+}-doped water-proof fluoro-aluminate glass fiber laser by employing a spherically aberrated intracavity lens with chromatic aberration [3].

S. Kurimura and T. Yagi (eds.), *Proceedings of Laser Display and Lighting Conference 2025*, Springer Proceedings in Physics 429,
https://doi.org/10.1007/978-981-96-9782-3_4

The generation of optical skyrmions in green, orange, red and deep-red from Pr^{3+} doped lasers with a wedge-plate shearing interferometer configuration has also demonstrated, though the direct generation of optical quasiparticles from laser sources as an eigen state is still challenging [4,5].

In this paper, we report on the multicolor (cyan, green, orange, red, and deep-red) visible structured light sources based on compact and robust diode-pumped Pr^{3+} doped lasers. We believe that these multicolor structured laser source enable the development of ultra-fine 2D/3D microfabrication, ultra-high resolution laser scanning microscopy, and advanced light-sheet imaging systems.

References

1. Rao, A.S., Miike, T., Miamoto, K., Omatsu, T.: Direct generation of 523 nm orbital Poincaré mode from a diode-pumped Pr^{3+}:$LiYF_4$ laser with an off-axis optical needle pumping geometry, Opt. Express 29 (19), 30409-30418 (2021).
2. Kerridge-Johns, W.R., Rao, A.S., Fujimoto, Y., Omatsu, T.: Red, orange, and dual wavelength vortex emission from Pr: WPFGF fiber laser using a microscope slide output coupler, Opt. Express 31 (10), 16607-16614 (2023).
3. Yoneda, Y., Rao, A.S., Fujimoto, Y., Miyamoto, K., Omatsu, T.: Direct generation of multicolor Bessel beams from a Pr^{3+}: WPFG fiber laser, Opt. Express 32 (6), 9011-9018 (2024).
4. Kerridge-Johns, W.R., Rao, A.S., Omatsu, T.: Optical skyrmion laser using a wedged output coupler, Optica 11 (6), 769-775 (2024).
5. Yoneda, Y., Allam, S.A., Kerridge-Johns, W.R., Fujimoto, Y., Omatsu, T.: Multi-Color Optical Quasiparticle Laser Source Formed of a Pr^{3+} doped fiber laser with a dual-output coupling geometry, Las. Photon. Rev., 2401403 (2024).

Mode-locked laser diode with highly-stacked InAs quantum dots

Kouichi Akahane*[1], Satoshi Yanase[1,2], Atsushi Matsumoto[1], Toshimasa Umezawa[1], Naokatsu Yamamoto[1], Tomohiro Maeda[1,2], and Hideyuki Sotobayashi[2]

[1] National Institute of Information and Communications Technology, 4-2-1, Nukuikitamachi, Koganei, Tokyo 184-8795, Japan
[2] Aoyama Gakuin University, 5-10-1, Fuchinobe, Chuo, Sagamihara, Kanagawa 252-5258, Japan
akahane@nict.go.jp

Abstract. Mode-locked laser diodes were fabricated with highly-stacked InAs quantum dots (QDs) using a strain compensation technique. Due to the use of high-density QDs, a wide-band gain was obtained, which resulted in a narrow pulse width and high repetition rate. Optical pulse trains of 20, 80, and 100 GHz with about 1ps pulse width were obtained depending on cavity length.

Keywords: quantum dot, semiconductor laser, mode-locked laser, molecular beam epitaxy

Semiconductor quantum dots (QDs) have various features to improve semiconductor lasers. In particular, self-assembled QDs are widely used to develop high-performance semiconductor lasers. Energy levels of QDs can be changed by changing their size, and wide band emission can thus be obtained in self-assembled QDs. However, a sufficient gain is required to achieve wideband lasing, for which high-density QDs are essential. To achieve wideband lasing, we have developed a technique to increase the QD density using strain compensation. In this paper, we introduce a method to obtain high-density QDs and the application of high-density QDs for mode-locked lasers. The density of our QDs is sufficiently high to deliver high gain and absorption in lasers. Additionally, it can lead to a decrease in the cavity length of mode-locked laser diodes. Therefore, we can obtain a high-repetition-rate optical pulse train.

The laser structure was grown on an n-type InP(311)B substrate by solid source molecular beam epitaxy. N-type and p-type InAlAs layers were used for the under and upper cladding layers, respectively. In the active layer, a 1-layer stacked structure of InAs QDs was grown with InGaAlAs strain compensation layers. Finally, a p-type InGaAs layer was grown as a contact layer. After the growth, a ridge waveguide laser structure was fabricated with ultraviolet (UV) lithography, dry etching, deposition of the SiO_2 isolation layer, and metal evaporation for the electrode. We adopted a two-section mode-locked laser diode structure. The saturable absorption (SA) section was fixed at 50 μm, and the gain section was varied from 340 to 1940 μm. The SA and gain sections were separated by a 10 μm gap. The fabricated QD mode-locked laser diodes

S. Kurimura and T. Yagi (eds.), *Proceedings of Laser Display and Lighting Conference 2025*, Springer Proceedings in Physics 429,
https://doi.org/10.1007/978-981-96-9782-3_5

were characterized by I-L measurement, optical and electrical spectrum measurement, and pulse width measurement with various injection currents of the gain section (I_g) and applied voltages of the SA (V_{sa}) section.

Figure 1 shows enlarged optical spectra for a cavity length of (a) 400, (b) 500, and (c) 2000 μm, with I_g and V_{sa} shown in the inset of each figure. All devices exhibited a wideband and clear longitudinal mode at around the 1550 nm telecom band. The separation of the longitudinal modes for the devices were (a) 100, (b) 80, and (c) 20 GHz, which were in good agreement with calculated values from the cavity length. In particular, a separation of around 100 GHz is suitable for WDM data communication, and more than 50 channels can be obtained for this QD mode-locked laser.

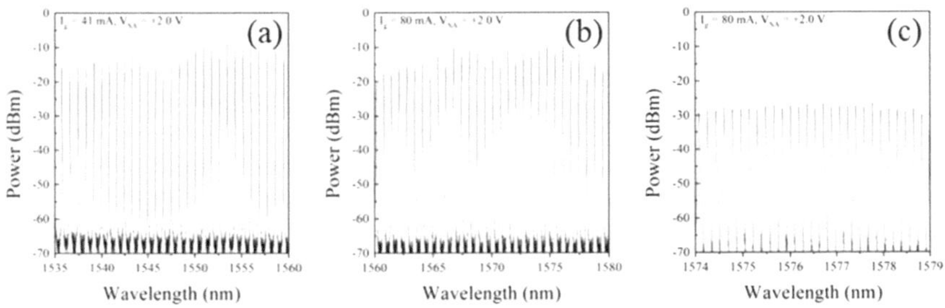

Fig. 1. Optical spectra of QD mode-locked laser with a cavity length of (a) 400, (b) 500, and (c) 2000 μm.

Figure 2 shows electrical spectra for the photoreceiver response with an output of a QD mode-locked laser with a 2000 μm cavity length. A clear peak can be observed at around 20 GHz, which indicates that the QD-mode-locked laser generates an optical pulse train. The frequencies in the electrical spectra show almost the same value of mode separation. The pulse width of these QD-mode-locked lasers measured by autocorrelation measurements with a Mikelson's interferometer is around 1 ps, which is attributed to QD wideband gain. Furthermore, in the I-L characteristics of these QD mode-locked lasers, the output power does not saturate, which is due to the use of high-density QDs. Therefore, we can obtain a higher repetition rate over 100 GHz with decreasing cavity length.

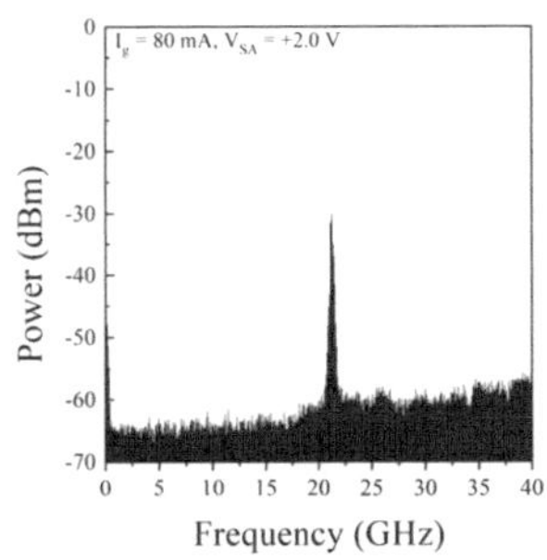

Fig. 2. Electrical spectrum of high-speed photoreceiver introducing output of QD-mode-locked laser.

InGaN-based Red Micro-Light-Emitting Diodes for Display Applications

Zhi Li[1], Brendan Roycroft[1], Changhao Li[1], Muhammet Genc[1], Abhinandan Hazarika[1], Bumjoon Kim[2], Soo Min Lee[2], Drew Hanser[2], Brian Corbett[1]

[1] Tyndall National Institute, University College Cork, Cork, T12 R5CP, Ireland
[2] Veeco Instruments Inc., Somerset, NJ 08873, USA
zhi.li@tyndall.ie

Abstract. The emerging InGaN-based red light-emitting diodes (LEDs) are promising candidates for display applications due to their potential high efficiency. In this work, we first investigated the electrical and optical properties of InGaN red LEDs on sapphire with various chip sizes (from 500×500 to 15×15 μm^2). By detailed characterization on the current-voltage measurement and current dependent light output power measurement, the negligible size-dependent effect was revealed. The suppressed surface recombination effect may be due to the effective carrier trapping effect by the V-pits, which prevents the carriers from flowing to the sidewalls. Furthermore, we developed releasing techniques for the InGaN red LEDs on silicon, enabling the integration via micro-transfer printing. The device performance after transfer printing was characterized in detail. The integration of red, green and blue InGaN-based micro-LEDs via micro-transfer printing were successfully demonstrated. This work paved the way for integrating InGaN red LEDs in the displays for higher performance.

Keywords: InGaN, Red LEDs, Surface Recombination, Displays, Transfer Printing.

1 Introduction

Micro-light-emitting diodes (micro-LEDs) have been a major development focus from both the academia and industry in the past decade, due to their huge potential in the display markets. Compared to conventional display technologies, such as liquid crystal display and organic LED display, micro-LED technologies exhibit superb advantages in the high brightness high contrast, low energy consumption and long lifetime [1]. One of the main challenges which hinders the further commercialization of this technology lies on the device efficiencies which drop significantly with the shrinkage of chip sizes [2]. Such size-dependent effect is associated with the surface recombination issues, which is especially serious in the traditional AlGaInP-based red LEDs.

To overcome the surface loss of the micro-LEDs, particularly for the red, several approaches [3] have been proposed and investigated, including surface passivation

S. Kurimura and T. Yagi (eds.), *Proceedings of Laser Display and Lighting Conference 2025*, Springer Proceedings in Physics 429,
https://doi.org/10.1007/978-981-96-9782-3_6

techniques, low-damage dry etching, and working on InGaN materials. It is known that the carrier localisation effect due to the non-homogenous indium distribution in InGaN quantum wells results in much shorter diffusion length than that in AlGaInP materials, which makes InGaN LEDs less sensitive to defects. Therefore, InGaN-based red LEDs have attracted much attention in recent years, due to their potential high efficiencies in micro-LEDs.

In this work, we studied the optical and electrical properties of InGaN-based red emitting LEDs grown on two different substrates, sapphire and silicon. For LED-on-Sapphire, the size-dependent behaviors were investigated; for LED-on-Si, the releasing of LEDs from substrate was investigated and transfer printing was demonstrated.

2 Experimental Results

The red LED structures were grown on 6-inch planar sapphire substrates and 8-inch silicon substrate with metal-organic chemical vapour deposition (MOCVD) respectively using a Veeco PROPEL® MOCVD system. The red LED was grown based on the V-pit structures, with the active regions consisting of a single InGaN (1 nm) blue quantum well (QW), 2 pairs of InGaN (2.5 nm) red-emitting QWs with each embedded in AlGaN (1.5 nm)/GaN (16 nm) barriers. For the basic device fabrication, the indium-tin-oxide (ITO) layers were used as the transparent electrodes on top of p-GaN, followed by the mesa definition using inductively coupled plasma etching. The sidewalls were passivated with SiO2 layers deposit by plasma-enhanced chemical vapor deposition and finally the big metal pads (Ti/Au) were deposited on both p and n sides for testing.

LED-on-Sapphire. The mesas (defined as chip sizes) were etched into square shapes with sizes from 500×500 to 15×15 μm^2. The optical images, current-voltage (I-V) characteristics, and power densities plots of fabricated red LEDs with various sizes were shown in Figure 1. Note that for each chip size, multiple devices (≥ 5) from various locations were tested and all the plots were displayed in the figure with the same color. For the I-V curves at the forward bias, all the plots were overlap with each other regardless of the chip sizes, even in the regions below the turn-on voltages. As the current densities in these low-voltage regions were associated with the sidewall leakage, such behavior indicates the negligible surface recombination loss in these InGaN-based red LEDs. Power density with the increasing current density were also plot, as shown in Figure 1(c). Except for the 500 μm chips which exhibited power density drops due to the current spreading issues, the other devices show very consistent behaviors. Although the 15 μm chips show noisy signals due to the very low power level, they still overlap much with other curves. Such size-insensitive behaviors further confirm the weak or negligible sidewall effects.

To understand the mechanisms of the suppressed surface recombination in these LEDs, the spatial electroluminescence images were taken and analyzed.

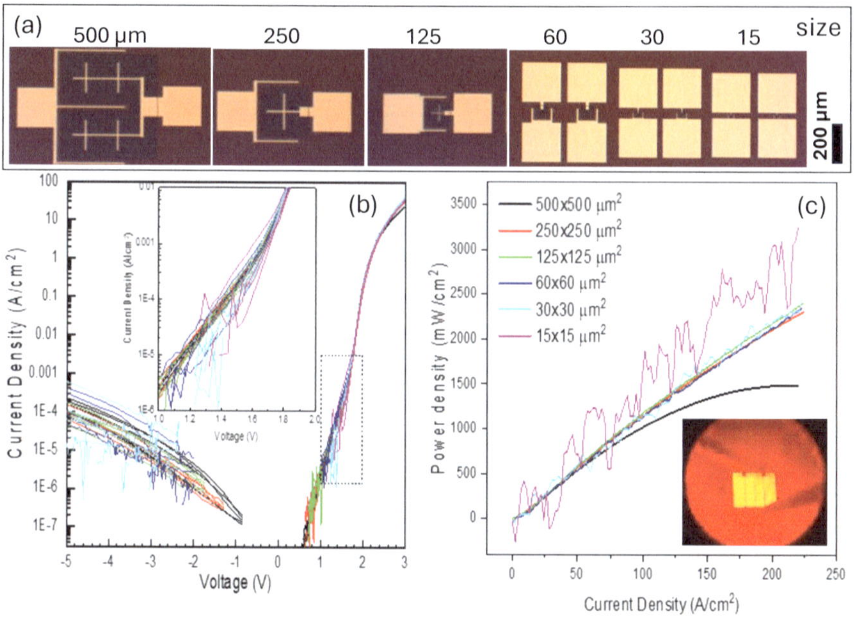

Fig. 1. (a) Optical images of InGaN red LED-on-Sapphire with the mesa sizes from 500 × 500 μm^2 to 15 × 15 μm^2. (b) Current density vs Voltage plots of these LEDs in log scale. The insert shows the enlarged region below the turn-on voltage. (c) Power density vs Current density plots of these LEDs. The insert shows the emission image of one such LED.

LED-on-Si. LEDs grown on substrate provide several advantages such as low cost, compatibility with the Si-based complementary metal-semiconductor (CMOS) process and easy releasing from the substrates. The LEDs with the size of 30×50 μm^2 were fabricated and anchored onto the Si substrate with SiN_x layers, as shown in Figure 2 (a)-(b). To release the devices from the substrate, the 5% tetramethylammonium hydroxide (TMAH) solution was used to etch Si at 75°C. Since the wet etch of (111) Si was anisotropic, the regions underneath the devices were etched much faster than the anchors [4]. The completely released LEDs were shown in Figure 2(c), where the devices were suspended on the surface and supported by the SiN_x tethers. Micro-transfer printing was then carried out, which picked up the devices and printed on a new target successfully, as shown in Figure 2(d).

The printed devices were connected with big metal pads after the printing process, and their I-V characteristics and spectral behaviors were tested. The blue and green InGaN-based counterparts were also printed and integrated with the red, demonstrating the RGB InGaN micro-LED integration.

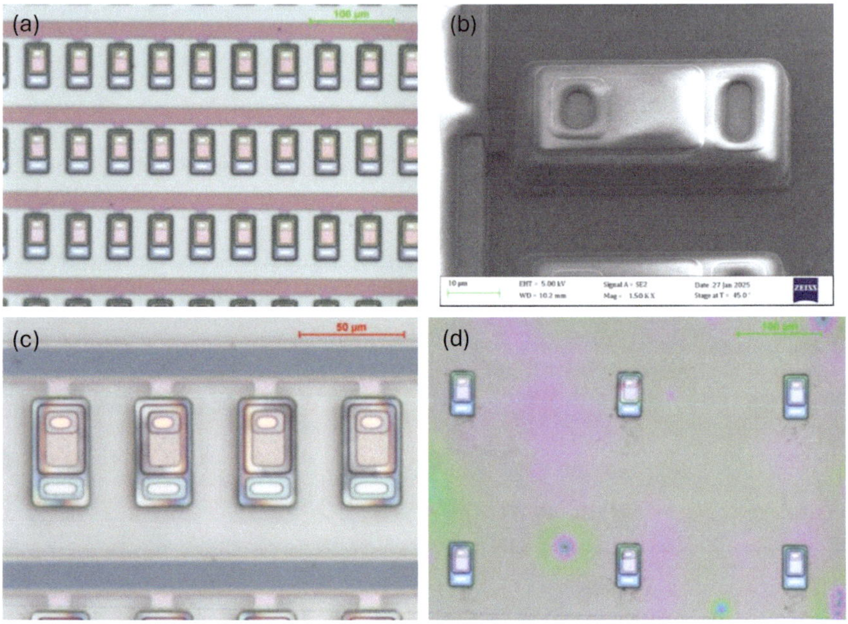

Fig. 2. (a) Fabricated InGaN-based red LED-on-Si prior to releasing etch. (b)The SEM image of a single LED before releasing. (c) Released (suspended) InGaN red LEDs on Si surface, still hooked to the anchor. (d) Printed red LED array on a new target wafer using micro-transfer printing technology.

Acknowledge. This work is funded by Research Ireland Pathway Program (SFI-IRC_22/PATH-S/10800) and Irish Photonics Integration Centre (IPIC) Program (SFI-12/RC/2276_P2_IPIC).

References

1. Parbrook, P. J.; Corbett, B.; Han, J.; Seong, T.; Amano, H. Micro-Light Emitting Diode: From Chips to Applications. Laser Photonics Rev., 15 (5), 2000133 (2021).
2. Li, Y.-Y.; Lin, F.-Z.; Chi, K.-L.; Weng, S.-Y.; Lee, G.-Y.; Kuo, H.-C.; Lin, C.-C. Analysis of Size-Dependent Quantum Efficiency in AlGaInP Micro–Light-Emitting Diodes With Consideration for Current Leakage. IEEE Photonics J. 14 (1), 1–7 (2022).
3. Liu, Z., Cao, H., Tang, X. *et al.* Advanced technologies in InGaN micro-LED fabrication to mitigate the sidewall effect. Light Sci Appl **14**, 64 (2025).
4. Shaban, Z.; Li, Z.; Roycroft, B.; Saei, M.; Mondal, T.; Corbett, B. Adv. Photonics Res. 3 (8), 2100312 (2022).

Polymer Optical Waveguide for Display Applications

Takaaki Ishigure[1][0000-0001-5155-1740]

[1] Keio University, Yokohama 223-8522, Japan
ishigure@appi.keio.ac.jp

Abstract. We present a unique polymer optical waveguide with circular cores fabricated by our proprietary technology, the Mosquito method. We demonstrate how to form 3D wiring patterns with perpendicularly bent cores with very small bending radii. We discuss the operability of polymer waveguides in the visible wavelength range for display applications.

Keywords: Polymer optical waveguide, the Mosquito method, 3D wiring.

1 Introduction

In the early 2000s, research and development of polymer optical waveguides were extensively carried out with the aim of using them for data transmission between the LCD screen and main board of flip phones, and for communication between the LSI and memory on the main board of supercomputers. In these applications, large-core multimode polymer waveguides were suitable with the combination of low-cost optical transceivers composed of VCSELs. Hence, at the time, polymer optical waveguides were expected to be a simple "replacement for electrical wiring," but the required specifications could be adequately met by wiring with one or two optical fibers or by improving electrical transmission technology, so planar optical waveguides with multiple transmission paths were not necessarily required. For this reason, the polymer optical waveguides that were being developed at the time were not put to practical use and did not form a large market. Since then, the momentum for research and development into polymer waveguides has been picking up again for the first time in about 20 years. In recent years, expectations have been rising for "single-mode" polymer optical waveguides as a means of transmitting signals from photonic integrated circuits mounted near switch chips (ASICs) in networks within large-scale data centers (co-package optics technology). Traditionally, both single-mode and even multimode polymer optical waveguides have been fabricated utilizing photolithography process. Therefore, the cross-section of the cores is to be square or rectangle. Meanwhile, we have proposed circular-core polymer optical waveguides enabling higher connectivity with optical fibers with circular cross-sectional cores and with VCSELs that emit circular beam.

In this paper, we present our proprietary technology named the Mosquito method to fabricate circular core polymer optical waveguides[1]. In addition, due to its unique fabrication step, 3-dimensional wiring is possible by the Mosquito method. These

S. Kurimura and T. Yagi (eds.), *Proceedings of Laser Display and Lighting Conference 2025*, Springer Proceedings in Physics 429,
https://doi.org/10.1007/978-981-96-9782-3_7

polymer optical waveguides have been primarily expected to be used for data transmission functions, but further applications in display technology are also discussed.

2 The Mosquito method

Fig. 1 shows the procedure of the Mosquito method. In the Mosquito method, first, a liquid state cladding monomer is coated on a substrate. Next, a viscous liquid-state core monomer is dispensed into the cladding monomer from a thin needle of a syringe while the needle scans horizontally and even vertically for drawing 3-D wiring patterns. Finally, the core and cladding monomers are simultaneously cured under UV exposure, followed by post baking to obtain the waveguides.

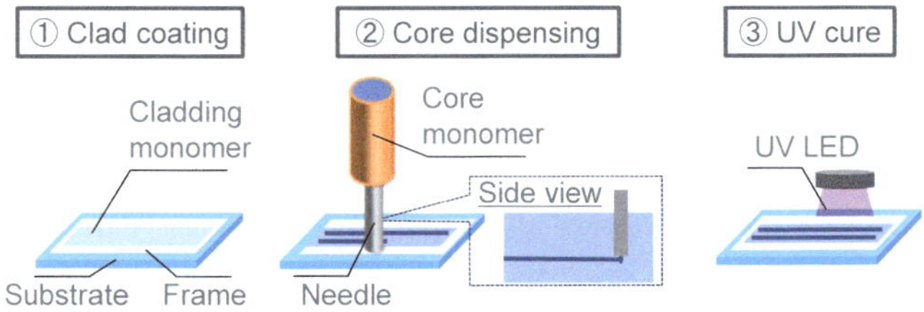

Fig. 1 Fabrication steps of polymer waveguide in the Mosquito method.

Here, we demonstrate how to apply the Mosquito method to form a 3D wiring pattern, i.e., a vertically 90-degree bent core waveguide. In step 2 of Fig. 1, we first place the tip of the needle directly above the VCSEL chip and start scanning toward the vertical direction, as shown in Figure 2(a). After a vertical scan of several millimeters long, the scan path of the needle is bent vertically 90 degrees with a radius of 1-2 mm. From the side-view photograph of the waveguide shown in Fig. 2(b), we confirm the vertically 90-degree bent core with a bending radius of 2.05 mm.

For display applications, propagation loss of polymer waveguides particularly at short wavelength (blue color) should be a key issue. Polymer materials with low absorption in such short wavelengths are highly anticipated.

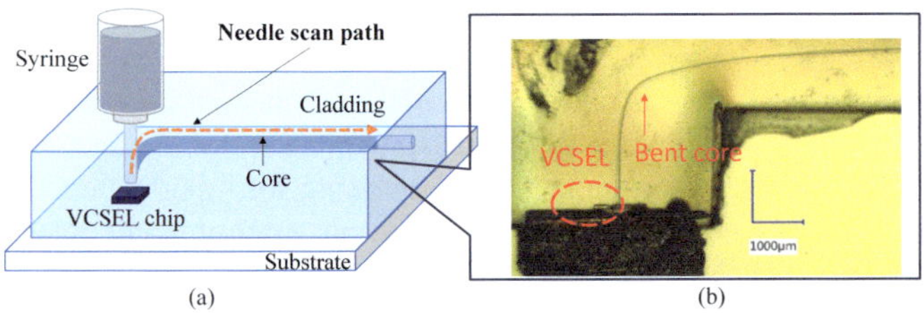

Fig. 2 Fabrication of vertically 90° bent core polymer waveguide.

References

1. Soma, K and Ishigure, T.: "Fabrication of a graded-index circular-core polymer parallel optical waveguide using a microdispenser for a high-density optical printed circuit board," IEEE J. Sel. Top. Quant. Electron., 19 (2), 360310 (2013)

Development of Microlens Array Mold and Periodic Submicron Structure Mold by Using Single Crystalline Diamond Tools

Hiroshi Saito[1[0000-0002-0738-7848]], Tsuneyuki Kobayashi[1], Yugo Sasaki[1], Satomu Kanazawa[2], Takako Matsumura[2], Kazumi Sawamura[2], Eiji Shamoto[3[0000-0001-7312-5478]] and Hirotsugu Yamamoto[4[0000-0003-1405-6657]]

[1] Yamagata Research Institute of Technology, 2-6-12 Matsuei, Yamagata-shi, Yamagata, Japan
saitohiroshi3@pref.yamagata.jp
[2] IMUZAK Inc., 2-6-12 Matsuei, Yamagata-shi, Yamagata, Japan
[3] Nagoya University, Furo-cho, Chikusa-ku, Nagoya-shi, Aichi, Japan
[4] Utsunomiya University, 7-1-2 Yoto, Utsunomiya-shi, Tochigi, Japan

Abstract. This study deals with the development of two optical components: a microlens array for air floating images and a periodic V-groove structure for anti-reflection. In the development of the microlens array, optical design and mold manufacturing methods are examined to simultaneously improve the resolution and the viewing angle of the floating image. Geometries of the microlens array are optimized using an optical design software CODE-V and the floating image is simulated using LightTools. Subsequently a cutting experiment of the microlens array mold with a single crystalline diamond tool is conducted using an ultraprecision 5-axis machine tool. An electroless Ni-P plating with a diagonal of 6.5 inches is used as a workpiece, and the changes in form error are evaluated. As a result, a resolution of MTF 0.5 at 1 cycle/mm and a viewing angle of 45 degrees is achieved. Changes in form error of the mold is 1.08 μm at most, that is the next challenge. To improve the uniformity of the form accuracy, a new method of semiconductor processing is also introduced. In the development of periodic V-grooves, a new cutting method called "Nano-Pecking" is applied to improve machining efficiency. A single crystalline diamond tool with a wedge angle of less than 50 degrees is vibrated by an ultrasonic oscillator. V-grooves can be machined at several tens of times larger pick feed than the ordinary method. Periodic V-groove structures with a pitch of 200 nm is successfully machined on a 20 mm square electroless Ni-P plated workpiece.

Keywords: Microlens Array, Anti-Reflection, Single Crystalline Diamond Tool, Electroless Ni-P Plating.

1 Introduction

Optical components such as microlens are used to control the refraction of light, and that requires both shape error and surface roughness to be sufficiently smaller than the

S. Kurimura and T. Yagi (eds.), *Proceedings of Laser Display and Lighting Conference 2025*, Springer Proceedings in Physics 429,
https://doi.org/10.1007/978-981-96-9782-3_8

wavelength of light. In addition, microstructures with a size less than the wavelength of light are attracting attention as a functional surface that reduces light reflection. Injection molding is a low-cost mass production method, and various studies have been reported on mold manufacturing for optical components. Photolithography is a common method in semiconductor processing for producing 2-D microstructures, which allows uniform and submicron-level processing over large areas [1, 2]. However, it is difficult to control depth and shape in etching complex shapes [3], and its application to curved surfaces is not widespread. Furthermore, the process is complex and must be carried out in a cleanroom, which increases manufacturing costs. A machining process with a single crystalline diamond tool, which has high-degree of freedom in shape, is more suitable for manufacturing microstructures with large irregularities or curved surfaces [4, 5]. On the other hand, the large machining area increases the tool wear and the small size of microstructures increases the machining time. Therefore, each process has its advantages and disadvantages, and there is a need for a method that can manufacture large-area structures with a high degree of freedom in shape and high precision, and a method that can manufacture submicron-order microstructures with high efficiency and low cost. This study describes the development of two optical components with feature sizes ranging from sub-millimeter to sub-micron levels, and demonstrates improvements in mold accuracy and machining efficiency.

The first optical component is a microlens array for air floating images. Real images can be formed in front of the microlens array, which is visible with the naked eye. It is expected to be used as a new interface by combining it with an IR sensor which can detect the position of a fingertip. However, the microlens array for air floating images has problems with a narrow viewing angle and low resolution. To overcome the problems, optical design of the microlens array is performed using optical design software CODE V and LightTools. A unique macro program of the CODE V is devised to optimize the microlens array geometry. LightTools is used to simulate the air floating images formed by the optimized microlens array. After the optical design, cutting experiments of the microlens array mold are performed using a single crystalline diamond tool, and the changes in form error are evaluated by using a laser probe 3D measuring equipment. Injection molding with the microlens array molds is also conducted to demonstrate that a real image can be formed by the developed microlens arrays. In addition to the mechanical mold manufacturing process, a new manufacturing method applying semiconductor process, which is under development by the authors, is also described to increase the uniformity of the mold accuracy.

The second optical component is a sub-micron periodic V-grooves for anti-reflection. The authors have developed a new cutting process called "Nano-Pecking" that enables highly efficient micro grooving by applying an ultrasonic vibration to the cutting tool. A special single crystalline diamond tool with a small wedge angle of less than 50 degrees is an important point to fabricate high aspect V-grooves. Fundamental cutting experiments on electroless Ni-P plating are conducted, and the quality of V-grooves are observed by using a scanning electron microscope (SEM). Finally, it is demonstrated that V-grooves with a pitch of 200 nm can be machined on a 20 mm square workpiece.

2 Experimental Method of Microlens Array

2.1 Optical Design of Microlens Array

The layout of the air floating image device is shown in Figure 1. A microlens array made by injection molding has the same convex lens array in a hexagonal arrangement on both sides. By placing a set of microlens arrays fixed with adhesive in front of an object, a real image is formed in front of the microlens arrays, which can be seen with the naked eye. In addition to the microlens array method, the micromirror method is also known as a floating image device [6]. Each method has its own characteristics, and the microlens method has advantages such as high brightness of floating images and low-cost manufacturing. However, the microlens array method has challenges to be solved, i.e. narrow viewing angle and low resolution of the floating image. To overcome these problems, optical design of the microlens array was carried out using an optical design software CODE V and LightTools provided by Synopsys, Inc. First, parameters such as curvature radius of the lens, lens pitch and thickness of the microlens array were determined by operating ray tracing in CODE V. Then another software LightTools was used to simulate how the floating image looks. These geometrical parameters have an influence on viewing angle and image resolution simultaneously: expanded viewing angle makes the light rays focus at multiple points, resulting in blurred image and low resolution. Lens pitch is also an important factor for the image resolution: the smaller the lens pitch, the higher the resolution. Based on these characteristics, the authors wrote a macro program on CODE V to find an optimized geometry of the microlens array. The viewing angle and the resolution of the floating image were calculated by ray trace diagram and MTF (Modulation transfer function) at 1 cycle/mm, respectively.

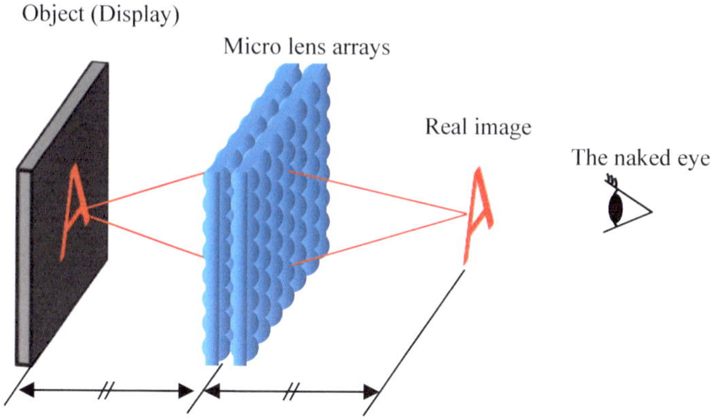

Fig. 1. Schematic illustration of the air floating image device.

2.2 Cutting Experiments of Microlens Array Molds

An ultraprecision 5-axis machine tool ROBONANO UiA made by FANUC Corporation was used in the cutting experiments of microlens array molds. A photograph of the experimental set-up and the schematic illustration of tool path are shown in Figures 2 and 3, respectively. The cutting tool was tilted at a 25 degree angle to avoid the use of the tool tip where the cutting speed becomes zero. Electroless Ni-P plating with a thickness of more than 200 μm was used as a workpiece material. After a roughing cycle with a sintered carbide ball end mill, a finishing cycle was performed using a single crystalline diamond tool with a nose radius equal to the curvature of the microlens. In order to suppress the wear of the diamond tool, another diamond tool with a different crystallographic direction was also prepared (hereinafter called "Wear resistant tool").

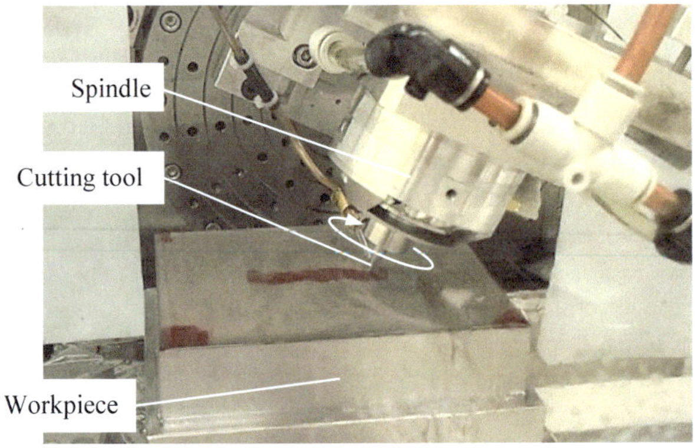

Fig. 2. Photograph of set-up for cutting experiment of micro lens array mold.

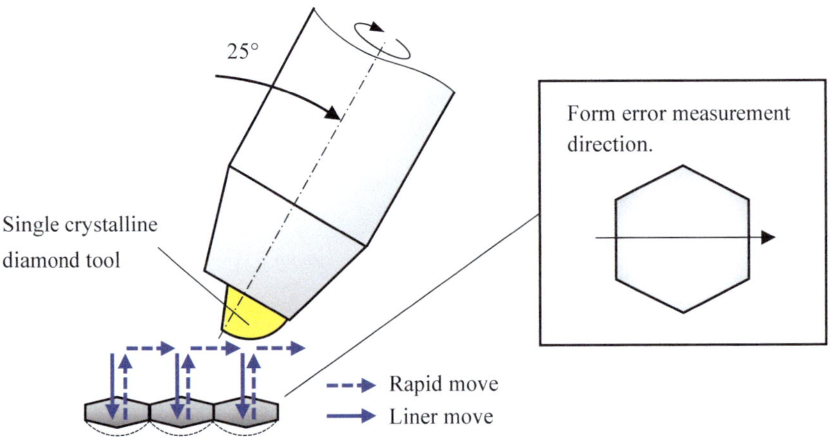

Fig. 3. Schematic illustration of tool path and form error measurement direction.

The details of the diamond tools are shown in Figure 4. The normal tool has a {001} face as the rake face and <001> direction is parallel to the rotation axis, whereas the wear resistant tool has a <001> direction inclined at 25 degrees to the rotation axis. The number of lenses machined was approximately 34000 in the case of the normal tool, and 68000 in the case of the wear resistant tool. Cutting conditions are summarized in Table 1. After the experiments, the form errors of microlenses were analyzed by using a laser probe 3D measuring equipment NH-3SP made by Mitaka Kohki Co., Ltd. In addition to the cutting experiments, injection molding was carried out using the machined molds to evaluate the quality of real floating images.

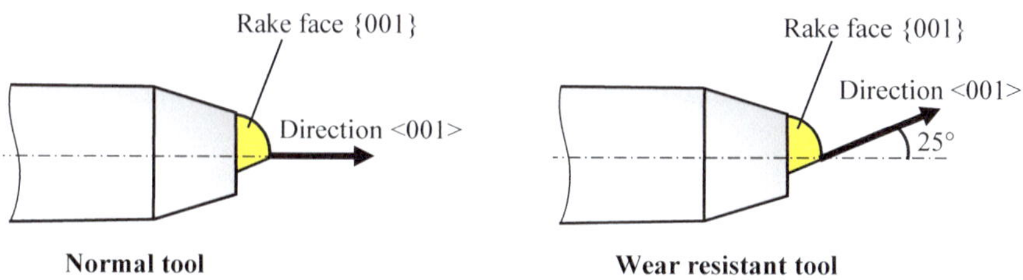

Fig. 4. Schematic illustrations of diamond crystallographic direction for normal and wear resistant tool.

Table 1. Machining conditions for microlens array molds.

Workpiece material	Electroless Ni-P plating
Workpiece size [mm]	134 * 102 (Diagonal 6.5 inches)
Number of workpieces	2 (for core and cavity)
Machine tool	FANUC Corporation, ROBONANO UiA
Cutting tool (Finishing cycle)	Single crystalline diamond tool
Finishing allowance [μm]	2-4
Revolution speed [min^{-1}]	23000
Feed speed [mm/min]	20
Cutting fluid	Oil-based cutting fluid
Number of machined lenses	Normal tool: Approximately 34000
	Wear resistant tool: Approximately 68000

3 Experimental Method of Periodic V-Groove Structures

3.1 A New Cutting Method "Nano-Pecking"

Figure 5 shows a schematic illustration of the developed cutting process "Nano-Pecking" compared with conventional cutting method "Shaping". In the case of shaping, one V-groove is machined in one reciprocating motion so that the pick feed must be same as the pitch of the V-grooves. The smaller the V-groove pitch, the longer the cutting time. On the other hand, in Nano-Pecking, a specially sharpened single crystalline diamond tool with a small wedge angle, e.g. less than 50 degrees, is vibrated with an ultrasonic oscillator in the intermediate direction between the nominal cutting and depth-of-cut directions. The V-groove is fabricated in a direction perpendicular to the nominal cutting direction. The detailed schematic illustration of Nano-Pecking is shown in Figure 6. The pitch of the V-grooves p_g can be expressed as a function of vibration frequency f and nominal cutting speed v:

$$p_g = v / f \quad (1)$$

Thus, pick feed becomes an independent variable from the pitch of the V-grooves, and it can be set tens or hundreds of times as large as that in the case of the shaping, which is one of the notable features of Nano-Pecking. The scallop height h, which represents the theoretical roughness perpendicular to the nominal cutting direction, can be approximately expressed [7] as a function of nose radius r and pick feed p_f:

$$h \approx p_f^2 / 8r \quad (2)$$

The pick feed p_f and the nose radius r should be selected so that the scallop height h becomes sufficiently small compared to the height of the V-grooves.

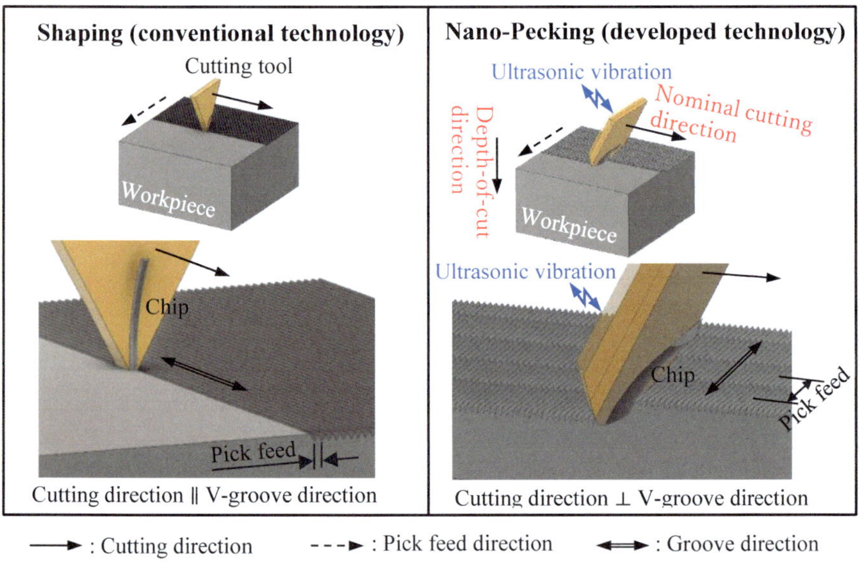

Fig. 5. Schematic illustrations of Nano-Pecking compared with conventional cutting technology.

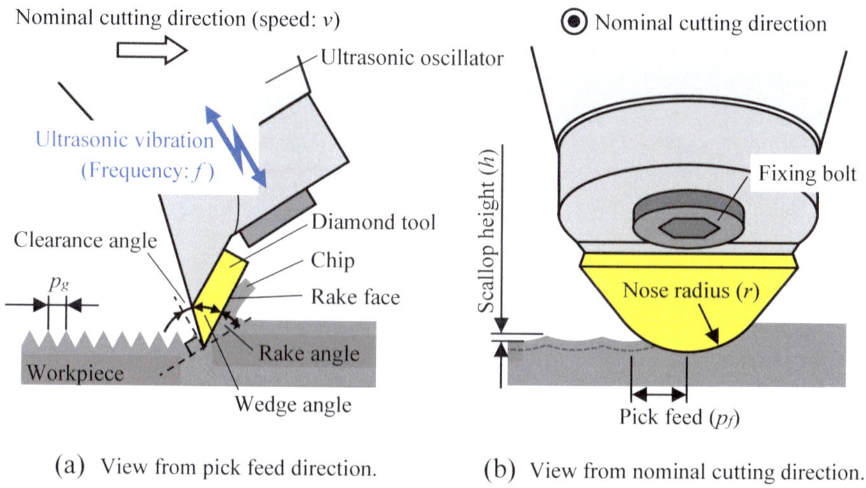

(a) View from pick feed direction. (b) View from nominal cutting direction.

Fig. 6. Detailed schematic illustrations of Nano-Pecking.

3.2 Nano-Pecking experiments

The experimental setup for Nano-Pecking experiment is shown in Figure 7. An ultraprecision machine tool Nagase Integrex, N2C-53US4N4, was equipped with an ultrasonic oscillator whose vibration frequency was approximately 20 kHz. A single crystalline diamond tool with a nose radius of 10 mm, rake angle of 35 degrees and clearance angle of 7 degrees was fixed on the tip of the oscillator. To fabricate V-grooves with a pitch of 200 nm, the cutting speed was calculated to be 240 mm/min using Equation 1. To machine V-grooves with high efficiency, the pick feed was set to 60 μm, and the scallop height was calculated as 45 nm using Equation 2. The aspect ratio of the V-grooves is estimated to be approximately 0.96 by drawing a cross-section diagram.

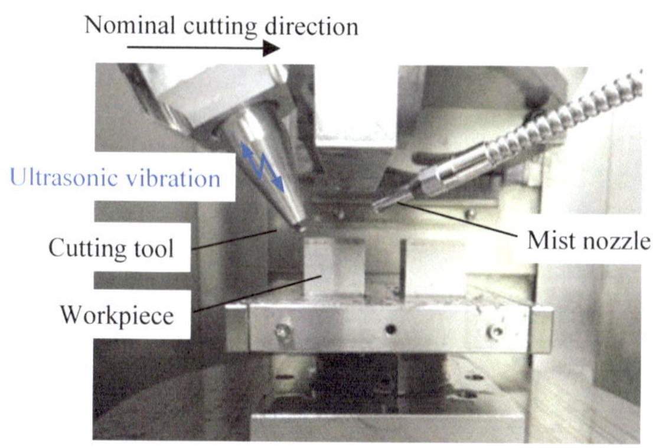

Fig. 7. Photograph of set-up for Nano-Pecking experiment.

After the experiment, the surface of the workpiece was observed by using a scanning electron microscope (SEM), JSM-7900F made by JEOL Ltd. Cutting conditions are summarized in Table 2.

Table 2. Cutting conditions for Nano-Pecking experiment.

Workpiece	Electroless Ni-P plating, 20*20 mm
Cutting tool	Single crystalline diamond tool
Nose radius [mm]	10
Rake angle [degree]	35
Clearance angle [degree]	7
Cutting speed [mm/min]	240
Depth of cut [μm]	Less than 3
Pick feed [μm]	60
Cutting fluid	Oil mist
Ultrasonic vibration frequency [kHz]	Approximately 20
Ultrasonic vibration amplitude [$\mu m_{p\text{-}p}$]	4

4 Results and Discussion

4.1 Optical Design Results of Microlens Array

Figure 8 shows the ray diagrams and MTF charts for the developed microlens array compared to the conventional one. The viewing angle corresponds to the angle made by the marginal rays at the output side, and was expanded from 35 degrees to 45 degrees

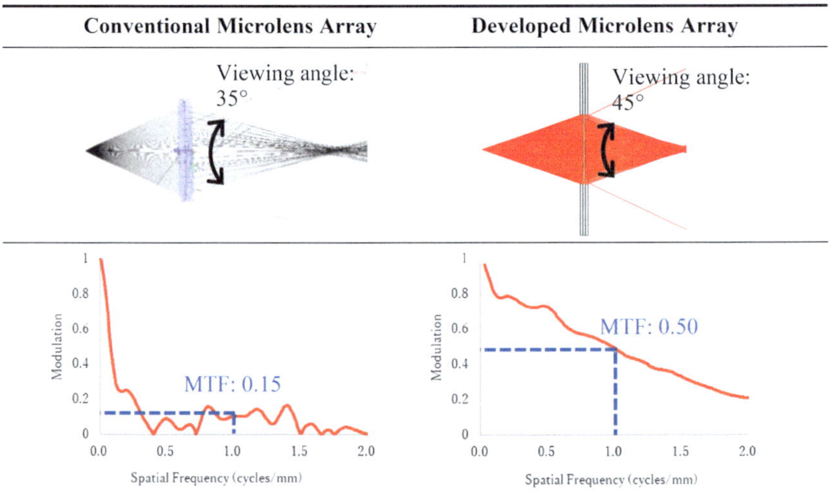

Fig. 8. Results of ray tracing and MTF calculation of microlens array.

by running a unique macro code on CODE V. Simultaneously, MTF at 1 cycle/mm was improved from 0.15 to 0.50, and the stray light at the output side was decreased in the case of the developed microlens array.

4.2 Results of cutting experiment of microlens array mold

Average values of form errors are summarized in Table 3. The initial values in the case of normal tools were 0.227 μm, then increased to 1.078 μm due to tool wear. On the other hand, the results in the case of wear resistant tools were 0.73 μm to 0.75 μm, although the number of machined lenses was twice as much as that in the case of the normal tools. Thus, tool wear was successfully suppressed by changing the crystallographic direction of diamond tools. However, the initial form error varies from 0.42 μm to 1.03 μm depending on the tool number, which implies the cutting-edge accuracy of the wear resistant tool is worse than that of the normal tool. The difficulty in sharpening the cutting edge of the wear resistant tool can affect the variation in the form accuracy.

Table 3. Average value of form errors against number of lenses machined.

Number of lenses machined	1	34000	68000
Normal tools [μm]	0.227	1.078	-
Wear resistant tools [μm]	0.728	0.744	0.747

Photographs of floating images and measured MTF values are shown in Figure 9. The floating image formed by the conventional microlens array was blurred and the measured MTF was 0.12 at 1 cycle/mm. In contrast, a clear floating image was obtained by the developed microlens array and the measured MTF was increased to 0.44. The difference between the calculated MTF values shown in Figure 8 and the measured values shown in Figure 9 are thought to be due to the form error of the mold. As described in the next chapter, the authors are developing a new manufacturing process to increase the uniformity of the mold accuracy.

Fig. 9. Photographs of floating images and its measured MTF values.

4.3 Mold manufacturing by applying semiconductor process

The authors are developing a new mold manufacturing process by applying semiconductor process to achieve higher uniformity of mold accuracy. The schematic illustration of the new process is shown in Figure 10. Photoresist is coated onto a glass substrate by spin coating to a thickness of 50 μm. One of the notable features of this process is the thick coating of photoresist. After prebaking, the microlens array pattern is exposed directly onto the photoresist using a scanning laser beam. This maskless lithography makes it possible to bypass the long photomask making process. A concave microlens array obtained by photoresist development is used as a mother mold for primary electroforming. Then the electroforming removed from the primary mother mold is used as a mother mold for secondary electroforming. Again, the nickel electroforming is removed and the outer periphery is machined to assemble into a mold base. This process will achieve a high uniformity in the microlens array mold.

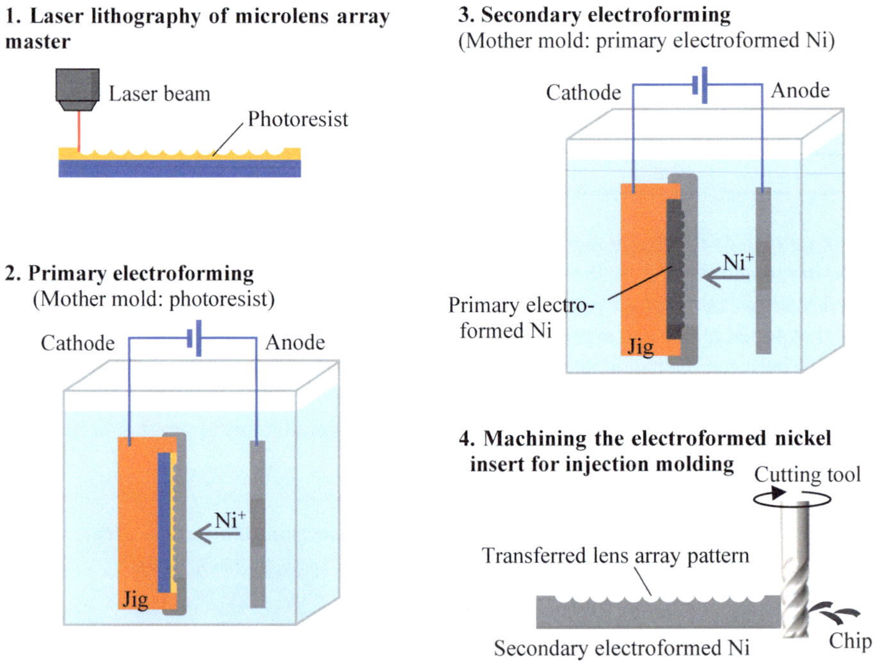

Fig. 10. Procedures for electroforming microlens array mold.

4.4 Result of cutting experiment on periodic V-groove structures

Figure 11 shows a photograph where V-grooves were machined by applying Nano-Pecking. The dark gray surface indicates that the reflection of light was reduced by the

V-grooves with a pitch of 200 nm. The SEM photographs of the V-grooves are shown in Figure 12. Uncut residues can be seen on the pick feed boundary area. In the Nano-Pecking process, it is difficult to align the vibration phase to the previous path, and the position of the V-grooves can shift in nominal cutting direction. This can be one of the reasons for the uncut residues. In other areas, the V-grooves with a pitch of 200 nm were successfully machined as shown in the enlarged photograph in Figure 12.

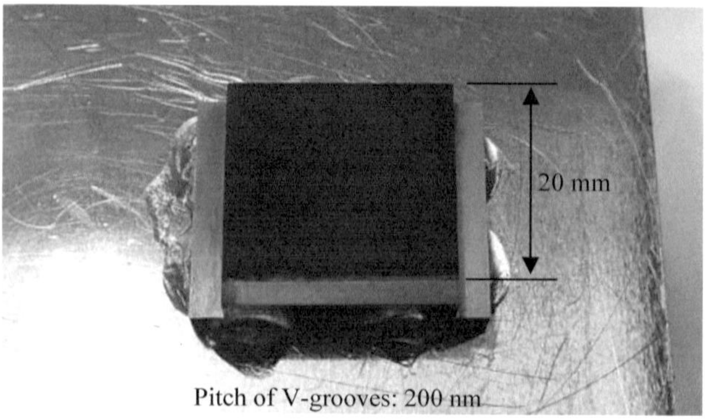

Fig. 11. Photograph of V-groove workpiece machined by Nano-Pecking.

Fig. 12. SEM photographs of workpiece machined by Nano-Pecking.

5 Conclusions

This research introduced the development of two types of optical components, i.e., a microlens array for air floating images and submicron periodic V-grooves for anti-reflection. Optical design of a microlens array was performed with CODE V and LightTools to improve the viewing angle and MTF of air floating images. After the optical design, cutting experiments of the mold were conducted with single crystalline diamond tools. Wear resistant tools which have a different crystallographic direction were also applied to the experiments. In the development of submicron periodic V-grooves, a new cutting technology "Nano-Pecking" was demonstrated using a 20 mm square electroless Ni-P plated workpiece. The results can be summarized below.

1. In the optical design of the microlens array, a viewing angle of 45 degrees and an MTF of 0.5 at 1 cycle/mm were achieved by applying optimization with a macro program in CODE V.
2. As a result of cutting experiments on microlens array molds, the average value of form error increased from 0.227 μm to 1.078 μm after machining 34000 lenses in the case of normal tools. On the other hand, no significant changes in form error were observed in the case of wear resistant tools, although the number of machined lenses was 68000.
3. A Nano-Pecking experiment was conducted and V-grooves with a pitch of 200 nm were successfully machined on a 20 mm square workpiece by applying a large pick feed of 60 μm.

Acknowledgments

The research was supported by "Improvement Support Operation for Strategic Foundational Technology" and "Growth-type Small and Medium-sized Enterprises Research and Development Support Operation" from Ministry of Economy, Trade and Industry. The authors would like to express their deepest appreciation for the supports.

References

1. Pan, C.T., Su, C.H.: Fabrication of gapless triangular micro-lens array. Sensors and Actuators 134(2), 631–640 (2007).
2. Tien, C.H., Chien, Y.E., Chiu, Y., Shieh, H.D.: Microlens array fabricated by excimer laser micromachining with gray-tone photolithography. Japanese Journal of Applied Physics 42(Part 1, No. 3), 1280–1283 (2003).
3. Kiihamäki, J., Kattelus, H., Karttunen, J., Franssila, S.: Depth and profile control in plasma etched MEMS structures. Sensors and Actuators A: Physical 82(1–3), 234–238 (2000).
4. Yan, J., Oowada, T., Zhou, T., Kuriyagawa, T.: Precision machining of microstructures on electroless-plated NiP surface for molding glass components. Journal of Materials Processing Technology 209(10), 4802–4808 (2009).
5. Kono, D., Nakade, S., Matsubara, A.: Evaluation of Surface Roughness in High-Speed Shaping of Ni-P. Key Engineering Materials 516, 125–129 (2012).
6. Maeda, Y.: Air Floating Image based on a Dihedral Corner Reflector Array. Proceedings of the 26th International Display Workshops 105–108 (2019).
7. Wang, Y., Shamoto, E.: Elliptical Vibration Cutting of Hardened Steel with Large Nose Radius Single Crystal Diamond Tool. International Journal of Automation Technology 8(6), 820–826 (2014).

Deformable Mirror Characterization for Efficient Fiber Coupling

Fergal Shevlin

Dyoptyka, Dublin, Ireland. fshevlin@dyoptyka.com

Abstract. Randomized divergence is considered stochastic and device characterization undertaken to estimate its maximum angular extent.

Keywords: deformable mirror, optical efficiency, speckle.

1 Introduction

Our company's high-frequency deformable mirror technology for speckle reduction is described in Fig. 1. Its randomized divergence has a small angular extent which allows optically efficient coupling into small core diameter optical fiber. For example, half-angle divergence of 1.5 deg can be coupled into 0.22 N.A. multimode optical fiber with $0.22/\sin 1.5 \approx 8\times$ demagnification which is sufficient for a $\varnothing 800\,\mu\text{m}$ spot on the mirror to be imaged into the entrance face of a fiber with core $\varnothing 100\,\mu\text{m}$, see [1].

Fig. 1. [Left] Random surface deformations are excited at frequencies up to 1.5 MHz in a 3 mm × 4.5 mm elliptical region. Reflectance $\geq$ 95% and damage threshold $> 1\,\text{W}\,\text{mm}^{-2}$ for wavelength-appropriate mirror coatings. [Center, left] Light from a $\varnothing 1$ mm beam incident to the mirror is diverged randomly within a small angular extent. [Center, right] Coupling lens images illuminated patch on mirror into entrance face of optical fiber. [Right] Exit face of $\varnothing 200\,\mu\text{m}$ 0.39 N.A. multimode fiber acquired with $20\,\mu\text{s}$ sensor exposure period; good homogeneity versus the typical rice grain pattern.

For optical efficiency and temporal stability, the coupling optical system should accommodate the maximum angular extent of randomized divergence. Due to the random nature of mirror surface deformations, as evidenced by their generation of uncorrelated speckle patterns, we consider angular extent as stochastic. See Fig.2 which shows how the randomized divergence of different $20\,\mu\text{s}$ sensor exposure periods varies significantly; as do tip and tilt angles which result in variation of position.

S. Kurimura and T. Yagi (eds.), *Proceedings of Laser Display and Lighting Conference 2025*, Springer Proceedings in Physics 429,
https://doi.org/10.1007/978-981-96-9782-3_9

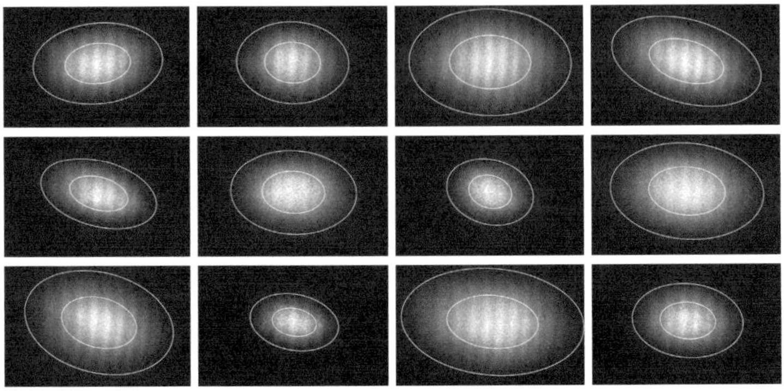

Fig. 2. Light from Thorlabs S4FC $\lambda = 520\,\text{nm}$ laser source, collimated into a ⌀500 µm beam, randomly diverged by mirror directly onto the 1920×1080 pixel sensor of a Basler Ace acA1920 camera operating at 1 Hz with a 20 µs exposure period. Interference fringes arise from its protective window. The contours of best-fit 2-D Gaussians are plotted at 1σ and 2σ distances.

2 Characterization

A deformable mirror was configured such that its half-angle divergence was approx. 1.5 deg when observed visually. Over 1000 images were acquired. For this first investigation, we assumed both divergence and position as 2-D Gaussian. Standard deviations and centroids are shown in Fig. 3. Taking the 2σ contour as the extent of randomized divergence and the 3σ contour as the extent of tip and tilt, the half-angle of maximum extent was calculated to be approx. 3 deg.

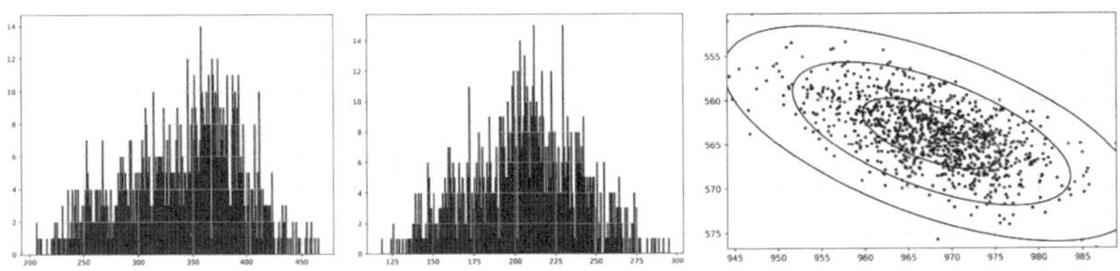

Fig. 3. [Left pair] Histograms of standard deviations along major and minor axes of best-fit 2-D Gaussians for 1000 images. [Right] Centroids of image regions. Best-fit 2-D Gaussian contours are plotted at 1σ, 2σ and 3σ distances.

References

1. Shevlin, F. High-Frequency Homogenization of Laser Illumination Through Stationary 0.22 N.A. Multimode Optical Fiber, Laser Display and Lighting Conference, Yokohama, Japan (2023.)

Holographic patterning of diffractive optical elements for beam shaping in lighting applications – a dual-hologram approach

Jorge Lasarte[1], Emma Branigan[1], Kevin Murphy[1], Izabela Naydenova[1], Jesús Atencia[2], Mª Victoria Collados[2] and Suzanne Martin[1].

[1] Technological University Dublin, School of Physics, Dublin, Ireland

[2] Engineering Research Institute of Aragon (I3A), Faculty of Science, Universidad de Zaragoza, Zaragoza, Spain

suzanne.martin@tudublin.ie

Abstract. The widespread use of Light Emitting Diodes (LEDs) in contemporary applications is a result of their many advantages such as cost-effectiveness, mass manufacturability, long lifespan, versatility, and lightweight construction, it has made them the preferred light source in many imaging lighting and projection systems. Holographic Optical Elements (HOEs) are optical elements fabricated using holographic techniques to pattern relatively large areas in low-cost optical films. They have potential in applications where lighting or displays would benefit from a shaped output that targets a specific audience location or illuminated zone. Bespoke elements can be mass produced with whole-field optical patterning using laser exposure energies of just a few milijoules (mJ), with the potential to reduce light pollution and increase energy efficiency. As such, HOEs present a promising approach to providing low-cost optical films for shaping LED sources.

In this paper, holographic recording is used to separately fabricate two HOEs which are then stacked to shape LED light. We report (1) a HOE combination that produces an on-axis collimated output from a divergent LED input beam and (2) a HOE combination that converts a diverging input beam into a structured beam determined by a Spatial Light Modulator (SLM). Output beams showing images and text highlight the flexibility that can be achieved to pattern and eventually shape the output from an LED source for bespoke lighting and display solutions. The dual-hologram approach could also simplify mass-production with the first hologram tailored to the LED source and the second to the desired output beam.

Keywords: On-axis, Collimated, HOE, LED, Image, Beam shaping.

S. Kurimura and T. Yagi (eds.), *Proceedings of Laser Display and Lighting Conference 2025*, Springer Proceedings in Physics 429,
https://doi.org/10.1007/978-981-96-9782-3_10

1 Introduction

The collimation of LED emissions is important for many applications requiring beam control, such as imaging systems, optical communications, bespoke lighting solutions, and augmented reality displays. Both diffractive and holographic optical element devices are commonly used to shape the output from laser sources [1-3] but the diverging cone of emitted light from an LED presents a more challenging problem. Like diffractive optical elements (DOEs), HOEs have advantages in terms of compactness and reduced weight over traditional optical components like lenses and mirrors but they also have the capacity for customization [4]. Volume HOEs could offer an efficient solution for steering and shaping light from LED sources because, in theory, 100% of the input beam can be redirected into a single diffracted output order. However, in order for volume HOEs to redirect, shape, and pattern light, the input beam must be Bragg-matched to the interference pattern that the HOE accepts [5]. Furthermore, the range of angles that the output beam can make, relative to the input beam, is limited due to the constraints imposed by the physical thickness of the recording medium with respect to the spatial period of the holographically recorded interference pattern [6]. These parameters determine whether the hologram does or does not operate in the volume regime, where all light is diverted into a single order. Multiple diffracted orders can occur for small diffraction angles and on-axis single holographic optical elements.

Two component, stacked HOEs have been used in laser systems to increase the image field [7] and production of on-axis volume holographic lenses [8,9,10] for laser sources, along with the correction of aberrations such as lateral chromatic dispersion [11]. Here we seek to adapt the approach for use with LED sources, stacking two HOEs to achieve an on-axis, collimated output from a divergent LED input source. The dual-hologram approach could also simplify mass-production because the first hologram is tailored to the LED source and the second to the desired output beam, so that as long as they are standardized, the components will be interchangeable. In this paper, we will combine HOEs in order to demonstrate solutions to three key issues:

1. Creating HOEs suitable for use with the highly diverging light emitted from LEDs.
2. Increasing the range of angles the output light can be redirected through, to include very small diffraction angles and on-axis collimation.
3. Flexible patterning of the output beam.

2 Materials and Methods

2.1 HOE stack to convert an on-axis diverging input beam to an on-axis patterned output

Bayfol HX200 [4] photopolymer film was used for recording. For the production of HOE 1, the central angle between the recording beams was set to either 24.3°, with a spatial frequency (SF) range of 475 - 1101 lines/mm, or 58°, with an SF range of 1400 – 2210 lines/mm, in air. A 20x microscope objective was placed in the object beam, 29 mm from the sample, to mimic the output from a highly divergent source [5]. The photopolymer sample was placed in the plane where the object and reference beams path overlapped, and the interference pattern was recorded with a total incident power on the sample of approximately 50 μW/cm^2, with an exposure time of 140 s.

HOE 2 was produced to collimate or pattern the off-axis input. The desired diffracted output pattern was introduced to the object beam using either a collimating lens or a spatial light modulator (SLM) to shape the wavefront. Once the photopolymer material has been exposed, the holographic element is fixed by exposing it to room light after the recording process and cannot be re-written afterward. To produce the desired pattern, the exposure times were optimized iteratively and were dependent on the pattern introduced by the SLM. For the recording of HOE 2 a total power of 45-66 μW/cm^2 and exposure times in the range 120-160 s were used. HOE 1 and HOE 2 (Figure 1) were stacked together and then aligned by hand in order to maximise the efficiency of the diffracted output.

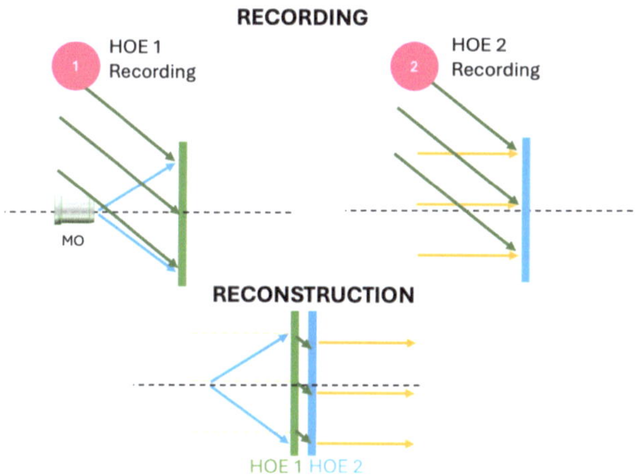

Fig. 1. The recording process used to fabricate the two HOEs and the reconstruction of the combined HOE stack. HOE 1 requires a diverging input to produce an output beam that is collimated and off-axis. HOE 2 requires an input beam that is collimated and off-axis and produces an output that is on-axis and contains the desired beam collimation or pattern.

3 Results

An angular selectivity analysis was performed on the on-axis collimated HOE 1 sample, and the results are shown in Figure 2. Here the HOE was rotated through a range of angles, while illuminated with a 532 nm source, and the intensity through the 0th order was captured with a photodiode. It is observed that the overall diffraction efficiency of the recorded HOE 1 is ≥ 80%. In order to demonstrate that the behavior of the recorded sample follows its design, three points were probed, the centre and the ± 6 mm locations, from which two peaks are obtained for each probed location, those corresponding to the input interbeam angles used to record the hologram. All three angular selectivity curves share the one being at the 58° mark corresponding to the off-axis collimated beam that was used for the construction of the hologram, and the other corresponding to the divergent source should be at 0° from the centre and vary equally towards the edges. As can be observed in Figure 2 the second peak is shifted by the same amount 9° for both probings and only differentiated by the sign, as expected.

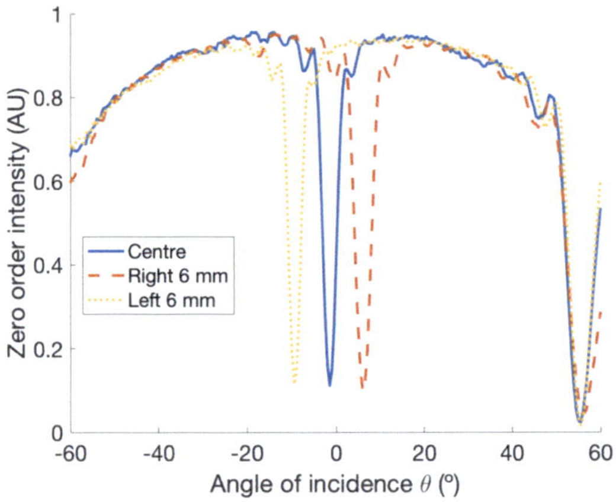

Fig. 2. The diffraction efficiency measured at three points for the on-axis collimated HOE 2 sample.

A slight drop in efficiency can be seen for the right 6 mm measurement but is expected since that area corresponds to the higher spatial frequency area where the lens has the highest angular selectivity. It is also worth noting that in this case the angle is 58° and such extreme angle probing on a lens is particularly challenging.

HOE 1 and 2 were then stacked and replayed with the diverging laser source and the LED, the output of which was either an on-axis patterned beam or reshaped, collimated beam (Figure 3).

HOE Stack	Laser	LED
Patterning element	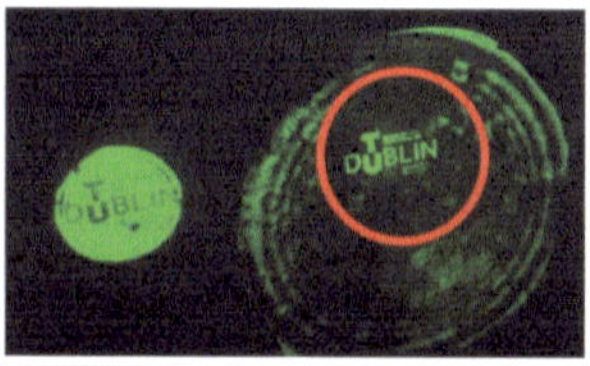	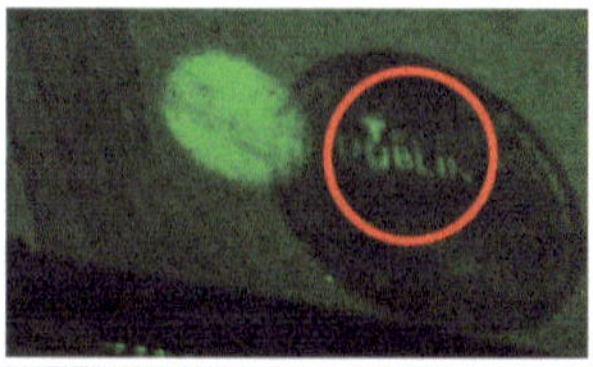
Collimating element	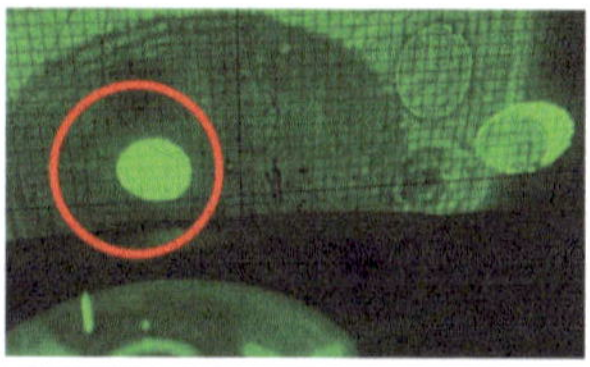	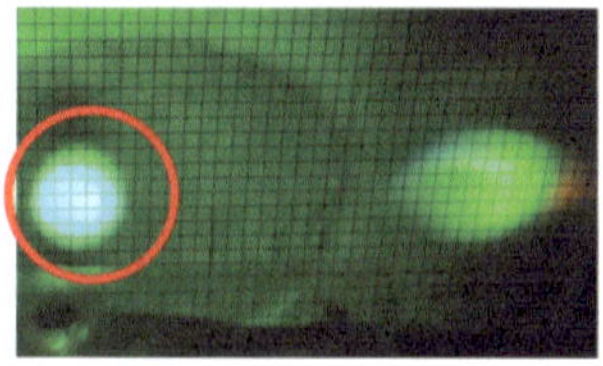

Figure 3. The diffracted output beams from the patterning and collimating HOE stacks under illumination from both a laser and LED source. Red circle around the on-axis stacked output.

Conclusion

A two-stacked HOEs approach to on-axis collimation or patterning of a diverging input light source was demonstrated. By implementing a stack of two HOEs, the range of working angles that the input beam can be redirected through has been increased. The HOEs were replayed both with a laser and an LED source, highlighting the flexibility and applicability of this HOE stack for bespoke lighting solutions. The recording beams of both HOEs in the stack can be tuned to accommodate different input LED sources and output patterns. In future, this work will be extended to include the shaping of diverging LED input sources to fully and efficiently redirect the source with minimal losses.

Acknowledgements

This work has been funded by Taighde Éireann - Research Ireland under grant number 20/FFP-P/8851 and by the Diputación General de Aragón-Fondo Social Europeo (TOL research group E44-23R). The authors thank the Physical to Life Sciences Hub at TU Dublin and Universidad de Zaragoza for the technical facilities and administrative support provided.

References

1. Vidal, E., et al.: Design and optimization of a collimating optical system for high divergence LED light sources. In: Current Developments in Lens Design and Optical Engineering X, vol. 7428. SPIE (2009).
2. Kozacki, T., et al.: LED near-eye holographic display with a large non-paraxial hologram generation. Opt. Express 30(24), 43551–43565 (2022).
3. Thiele, S., Gissibl, T., Giessen, H., Herkommer, A. M.: Ultra-compact on-chip LED collimation optics by 3D femtosecond direct laser writing. Opt. Lett. 41, 3029–3032 (2016).
4. Bruder, F. K., Fäcke, T., Rölle, T. "The Chemistry and Physics of Bayfol® HX Film Holographic Photopolymer." *Polymers (Basel)* 9, no. 10 (2017): 472. doi: 10.3390/polym9100472. PMID: 30965774; PMCID: PMC6418958.
5. Lasarte, J., Murphy, K., Naydenova, I., Atencia, J., Collados, M. V., Martin, S. "Modeling holographic optical element performance with an extended source; experimental investigation using misaligned point sources." *J. Opt. Microsystems*4, no. 1 (2024): 014002.
6. Murray, M., Naydenova, I., Martin, S. "Review of recent advances in photosensitive polymer materials and requirements for transmission Diffractive Optical Elements for LED light sources." *Opt. Mater. Express* (2023).
7. Chen, E., et al.: Broadband beam collimation metasurface for full-color micro-LED displays. Opt. Express 32(6), 10252–10264 (2024).
8. Collados, M. V., López, A. M., Atencia, J., Quintanilla, M.: Partitioned-field holographic lenses composed of three noncentered uniaxial systems. Appl. Opt. 42(32), 6445–6451 (2003).
9. Latta, J. N.: Computer-based analysis of hologram imagery and aberrations. I. Hologram types and their nonchromatic aberrations. Appl. Opt. 10(3), 599–608 (1971).
10. Keshri, S., et al.: Stacked volume holographic gratings for extending the operational wavelength range in LED and solar applications. Appl. Opt. 59(8), 2569–2579 (2020).
11. Martin, S., Akbari, H., Keshri, S., Bade, D., Naydenova, I., Murphy, K., Toal, V.: Holographically recorded low spatial frequency volume Bragg gratings and holographic optical elements. Hologr. Mater. Opt. Syst. 4, 73–98 (2017).
12. Latta, J. N.: Analysis of multiple hologram optical elements with low dispersion and low aberrations. Appl. Opt. 11, 1686–1696 (1972).

MEMS-based Time, Wavelength, and Diffraction Multiplexed Image Transfer for Display Engine

Yuzuru Takashima[1], Chuan Luo[1], Gregory Nero[1], Yexin Pei[1], Tianyao Zhang[1], Xianyue Deng[1], Ted Lee[1], and Parker Liu[1]

[1] Wyant College of Optical Sciences, University of Arizona, 1630 E. University Blvd. Tucson AZ 85721 USA
ytakashima@optics.arizona.edu

Abstract. MEMS (Micro Electro Mechanical System) spatial light modulators (SLMs) enable a flexible and fast control of light that opens pathway to compact near-to-eye display engine. We introduce novel use cases of the MEMS SLMs for time-of-flight lidar and AR projection engine that enables wide field-of-view (FOV) image transfer beyond FOV of image guiding device.

Keywords: MEMS Spatial light modulators, image steering, field-of-view expansion

1 Introduction

Near-to-eye display (NED) system, or "smart glasses" is an image data transfer system from micro display device to the human retina. The optical data channel from the micro display to retina consists of a projection engine, image transfer optics such as a Total Internal Reflection (TIR) image guiding device having input- and output-couplers while employing pupil expansion. The TIR-image guide and the free-space couplers accommodate the field-of-view (FOV) of over 30 degrees in diagonal as well as see-through images. One of the challenges in the NED is limited FOV and trade-off between FOV, image resolution, and panel size and pixel count. Since the number of pixels is a limited resource, to increase the FOV, image resolution is sacrificed. With wide-FOV and low-resolution images generated, system FOV is limited by FOV of the TIR image guide which is limited by refractive index of the TIR image guide. The resolution can be increased by increasing the number of pixels though the approach requires more volume for micro displays that ends with increased volume of optical engine.

One of the approaches to the trilemma is decoupling of FOV from pixel counts by time multiplexing and scanning of images. For image steering, we employ MEMS-based SLM, Texas Instruments Digital Micromirror Device (DMD) with a new pulsed illumination scheme. With DMD as a reflective MEMS SLM, we employ wavelength multiplexed image transfer that increases angular bandwidth of the TIR image guide by factor of 2 without increasing the RI of the image guide.

S. Kurimura and T. Yagi (eds.), *Proceedings of Laser Display and Lighting Conference 2025*, Springer Proceedings in Physics 429,
https://doi.org/10.1007/978-981-96-9782-3_11

2 Image steering by MEMS SLM

As a MEMS-SLM, we employ Texas Instruments Digital Micromirror Device (TI-DMD). Conventionally, DMD has been used as a binary display device that modulates the illumination light by on-and off-state of the MEMS mirror in tilt motion.
The TI-DMD consists of an array of micro mirrors that tilt in ±12 degrees that correspond to on- and off-pixel states, respectively [1]. The on- off-states of mirrors redirect incoming light to ±24 degrees, respectively where light reflected from on-pixel is captured by entrance pupil of the projection lens, whereas a light from off-states is redirected towards a beam dump. In between the on- and off-states, the tilt angle of mirror array continuously varies from +12 to −12 degrees. The transitional time between the on- and off-states is in the order of several microseconds and is known as crossover time. During the crossover time, the mirror array changes its tilt angle continuously. The key idea of the Angular and Spatial Light Modulation (ASLM) is accessing the transitional state of the mirror by using a pulsed laser having substantially shorter pulse width compared to the mirror transition time. The timing of the pulsed laser to the mirror transition adds phase tilt that diffracts beam and image into one of the diffraction orders with high efficiency. Figure 1b shows a phase profile of DMD mirror array while mirrors are in transition [2-5].

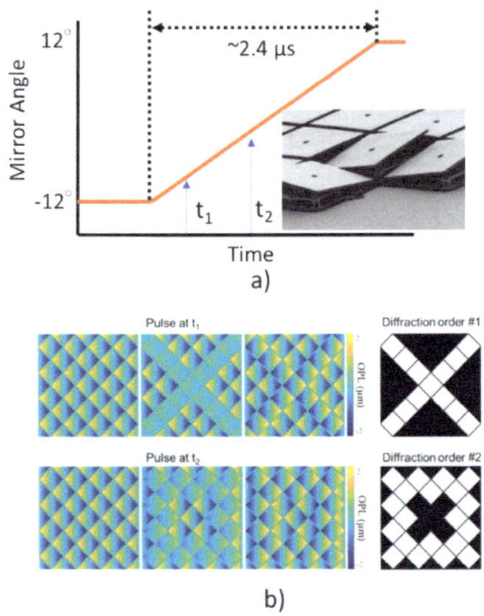

Fig. 1. Phase profile if DMD mirror array when (a) all the mirrors are actuated, (b) pattern is displayed, at timings t1 and t2 while mirrors are in motion [4].

3 Time and Wavelength Multiplexed Image Transfer via Angular Band-limited Image Guide

The image steering generates wide FOV and high-resolution image by time-multiplexed image steering over the FOV. As a next challenge, the wide-FOV (large angular bandwidth) image must be transferred from micro display to eye via angularly bandlimited TIR image guide or free-space optical combiner. Figure 2 shows a concept of time and wavelength multiplexed image transfer that enables wide FOV image transfer without using high index (n=2.0) TIR image guide. First, we use two laser sources, wavelengths λ1 and λ2. On micro display, the total FOV is divided into sub FOVs, a central FOV, and two peripheral FOVs for which wavelengths λ1 and λ2 are assigned, respectively. The total FOV is halved and transferred via band limited TIR image guide. Once the image is transferred, de-multiplexing is employed by de-multiplexing film layer that sorts three images in the original order. Finally, the original image is reconstructed and perceived by the human eye.

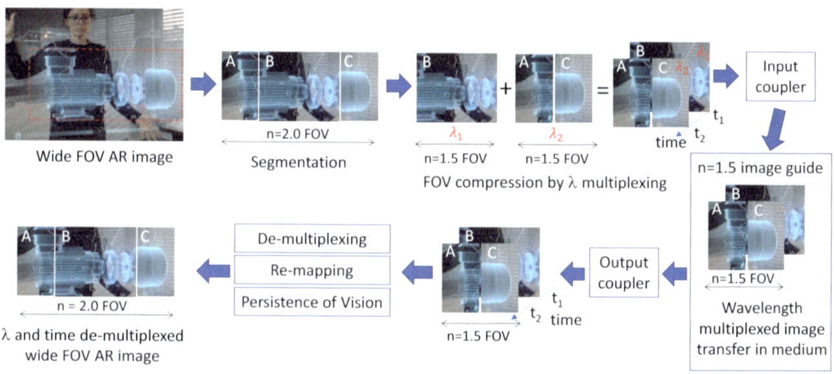

Fig. 2. A concept employing time and wavelength multiplexed image transfer. The multiplexing and de-multiplexing enables transfer wide FOV, i.e., 60 degrees via angularly band limited optical combiner, i.e., 30 degrees of FOV with n=1.5 RI TIR image guide [6, 7].

4 Experimental Results

Figure 3 shows a time and wavelength multiplexed image transfer via band limited optical channel. We formed the central image (30 degrees FOV) and peripheral images (15 degree FOV each). Central image is projected and captured by camera. The peripheral image is then displayed and captured by the camera again. Note that the only parameter in the two capturing processes is wavelength. The central and peripheral images are sorted to the intended position by stack of volume holographic gratings. The time wavelength multiplexing is further hybridized with diffractive multiplexing. Figure 4 shows the experimental results. The 30 deg FOV images are further divided into ten 6 degree sub FOVs (Fig. 3a). For each of the sub FOV images, wavelength and diffraction orders are assigned. For peripheral and central images, wavelength of 638nm and 658nm are assigned, respectively (Fig. 4a). The peripheral ana central image assigned to 638nm and 658nm is further segmented (Fig. 4b, c) and are diffractively steered by DMD and forms continuous image (Fig. 4d). In this manner, one fifth of portion of DMD that corresponds to 6 degrees of FOV is used to generate a wide FOV 60 degrees of image while transferring image beyond angular bandwidth of the image transfer device with 30 degree of FOV.

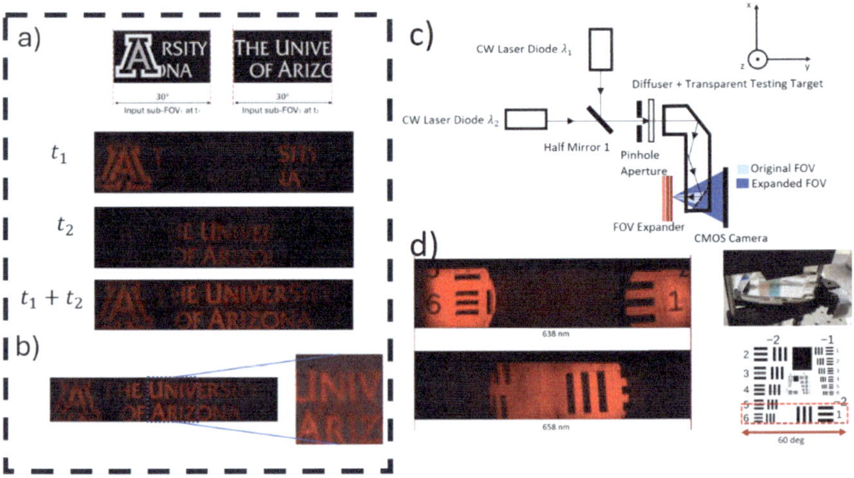

Fig. 3. Image transfer demo employing time and wavelength multiplexing. The 60 degree image is divided into two 30 deg FOV images, central and peripheral images. Those images are time multiplexed (Fig. 3a), transferred over FOV limited (30 degrees) TIR image guide (Fig. 3c). The images are demultiplexed by FOV expander (Volume Hologram Stacks), and 60 degree continuous image is formed (Fig. 3b) [5].

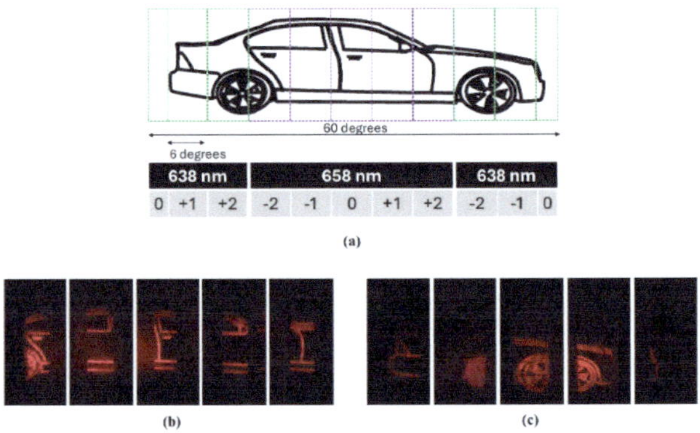

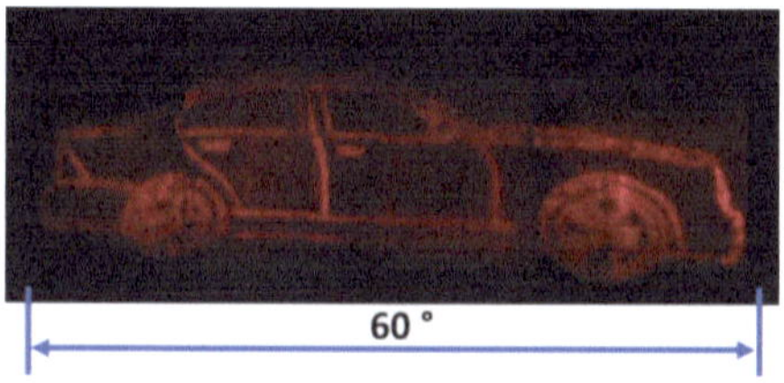

(d)

Fig. 4. Image transfer demo employing time, diffraction, and wavelength multiplexing. a) The total of 60 degree FOV images are divided into 10 sub images. For peripheral and central images, 638nm and 658nm wavelength are assigned (Fig. 4a). Each of the 6 degree images are time multiplexed while diffractively steered by DMD (Fig 4b, c). The ten elemental images are re-sorted by volume holographic grating and forms 60 degree continuous image. Note that the FOV of the image guiding device is 30 degrees [6].

5 Conclusions

Diffractive image steering by MEMS display device, Texas Instruments Digital Micromirror Device (DMD) solves common challenges, coupling of FOV, resolution, number of pixels, and form factor originated from conservation of Etendue. The time- and wavelength multiplexing enables image transfer with angular bandwidth that exceeds physically limited angular bandwidth of image transfer device, i.e., TIR image guide commonly employed in Near-to-eye (NED) displays. The fast framerate of MEMS SLM, diversity in wavelength and polarization uniquely positions MEMS SLM in NED display engine that solves challenges due to coupling of FOV, resolution and optics volume, by employing newly developed pulsed laser illumination the turns binary DMD into programable image steering spatial light modulator.

Acknowledgment: A part of the research is supported by Texas Instruments, and TxACE, Semiconductor Research Corporation.

References

[1] L. J. Hornbeck, "From cathode rays to digital micromirrors: a history of electronic projection display", TI Tech. J. 15, 7–46 (1998).
[2] B. Smith, B. Hellman, A. Gin, A. Espinoza, and Y. Takashima, "Single Chip Lidar with Discrete Beam Steering by Digital Micromirror Device", Optics Express, Vol. 25, Issue 13, pp. 14732-14745 (2017).
[3] B. Hellman and Y. Takashima, "Angular and spatial light modulation by single digital micromirror device for multi-image output and nearly doubled étendue," Opt. Express 27, 21477-21496 (2019).

[4] B. Hellman, Ted Lee, Y. Takashima, "Gigapixel and 1440-perspective extended-angle display by megapixel MEMS-SLM," Optics Letters 45.18, 5016-019 (2020).

[5] G. Nero, Y. Pei, Y. Zhang, X. Deng, C-I Tang, P. Liu, T. L-T Lee, M. Lunin, and Y. Takashima. "Field-of-view expansion via diffractive image steering and prism array", Proc. SPIE, 12449-117 (2023).

[6] Pengyu Liu, Chuan Luo, Ted L. Lee, Gregory Nero, Tianyao Zhang, Yexin Pei, Xianyue Deng, and Yuzuru Takashima, "Expanding field-of-view of near-to-eye display by hybrid angular and wavelength multiplexing," Opt. Express 33, 1466-1477 (2025).

[7] Lee, Ted; Liu, Pengyu; Luo, Chuan; Zhang, Tianyao; Pei, Yexin; Takashima, Yuzuru "Image Steering with Multi-domain Field of View Expansion by MEMS SLM for Near-to-eye AR Display Engines," Accepted for Journal of the Society for Information Display (2025)

Smart Contact Lenses

Dieter Cuypers[1] [0000-0003-0550-6273] and Herbert De Smet[1] [0000-0001-9640-8454]

[1] Centre for Microsystems Technology, imec and Ghent University, Technologiepark 126, B-9052 Ghent, Belgium
dieter.cuypers@ugent.be

Abstract. The technology track of smart contact lenses is introduced and discussed.

Keywords: Contact lens, liquid crystal lens.

1 Introduction

Smart contact lenses (SCL) are devices based on the standard contact lens substrate that offer additional functionality, unrelated to the primary, original function of vision correction. As such, this broad definition encompasses a variety of devices, ranging from sensing aids that remain in the medical or ophthalmological sphere to on-the-eye augmented reality displays.

From the viewpoint of the display industry, a first useful distinction can be made between image-forming and non-image forming SCLs. The latter comprise both the sensing and drug delivery devices which obviously have no display functionality at all, but also contact lenses intended as medical devices that do use display technologies, however only to assist or improve the viewing of the external world. Image forming SCLs have true display functionality, in that they seek to generate some form of imagery in the eye unrelated to the external world.

Serious research efforts have been made in both categories [1-7] but lasting products are yet to be produced, particularly for contact lens displays. In this talk, some technological aspects, particularly in relation to the image forming types will be discussed.

2 SCL concept

A conceptual drawing of a smart contact lens is shown in **Fig. 1**. Not all components need always be present, but in general there will be a controller chip, an energy storage and / or energy generating component, a communication device with the outside world such as an antenna and an active component. This active component is often electro-optic (like a display), but can also be a (bio-)sensor. All of this must be incorporated in

S. Kurimura and T. Yagi (eds.), *Proceedings of Laser Display and Lighting Conference 2025*, Springer Proceedings in Physics 429,
https://doi.org/10.1007/978-981-96-9782-3_12

the dimensional constraints of a contact lens, leaving enough transparency for the outside scenery to be perceived. The exposed outer materials should of course be bio-compatible.

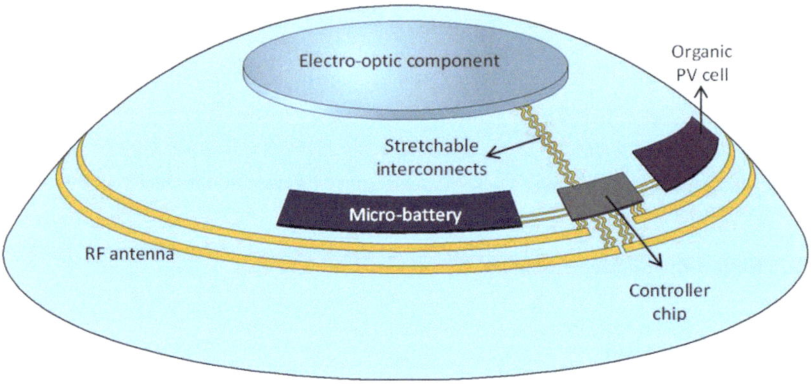

Fig. 1. A generic representation of a smart contact lens with display functionality, featuring the components needed for independent operation (electro-optic component, control chip, battery, charging cell and antenna).

References

1. Lingley, A R: A single-pixel wireless contact lens display. J. Micromech. Microeng. 21 125014 (2011)
2. Liao, Y.-T.: A 3-μW CMOS Glucose Sensor for Wireless Contact-Lens Tear Glucose Monitoring. IEEE J. Solid-State Circuits 47, 335 (2012)
3. Bailey, J.: Design considerations for liquid crystal contact lenses. J. Phys. D: Appl. Phys. 50 485401 (2017)
4. Nasreldin, M.: Flexible Micro-Battery for Powering Smart Contact Lens. Sensors 19, 2062 (2019)
5. Park, J.: Printing of wirelessly rechargeable solid-state supercapacitors for soft, smart contact lenses with continuous operations. Sci. Adv. 5(12), (2019)
6. Lan, S.: Metasurfaces for Near-Eye Augmented Reality. ACS Photonics 6, 864 (2019)
7. Sano J.: Holographic contact lens display that provides focusable images for eyes. Opt Express 29, 10568 (2021)

Visible near-to-eye display technology based on nonlinear frequency conversion

Peter G.R. Smith[1][0000-0003-0319-718X] Goronwy L. Tawy[1] [0000-0003-3508-3507] Rex H.S. Bannerman[1] [0000-0002-1428-6308] and Corin B.E. Gawith[1] [0000-0002-3502-3558]

[1] Optoelectronics Research Centre, University of Southampton, Highfield, Southampton, Hampshire, SO17 1BJ, UK

peter.smith@soton.ac.uk

Abstract. We present a new approach to near-to-eye displays based on the use of non-colinear wave-mixing in a nonlinear lens material. Conventional approaches to near-to-eye displays – used in Augmented Reality (AR) use linear optics, and this creates challenges with transparency, scatter and eye-glow, with creating 3d wavefront effects and brightness. Our new approach combines two infrared beams to create an exit-pupil that sends light to the viewer's eye. This approach allows us to create a fully transparent display, with high brightness, high optical efficiency, and the ability to fully control wavefront so as to create a virtual image at any desired distance. The human eye is extremely sensitive to light, and so we need only generate microwatts of light to create a bright image. We can create these outputs used pulsed microchip lasers which offer kW peak powers at mW level average powers.

Keywords: Augmented reality, Nonlinear Optics, Metaverse.

1 Background

1.1 Close-to-eye displays

Head-up-display technology has been used for applications such as fast-jet pilots for decades, providing visual information for the pilot which appears in their visual field. This concept has increasingly become reality for ordinary users – for example the well-known Google Glass initiative, and more recently many commercial products and demo devices from the Big Tech companies. Various approaches have been used including laser scanning, birdbath optics, diffractive/waveguides, metasurfaces, micromirrors, etc. An excellent recent review paper [1] describes each of these approaches and compares their performance. However, despite billions of dollars of investment in R&D none of these approaches has yet fulfilled the full range of required properties.

Close-to-eye displays are generally either virtual reality (VR) or augmented reality (AR) with the key difference being that in VR the headset is enclosed and all of the light coming to the eye is generated in the display itself. The display is thus completely blocking to light – and this means that the favored approach is to place an emissive

S. Kurimura and T. Yagi (eds.), *Proceedings of Laser Display and Lighting Conference 2025*, Springer Proceedings in Physics 429,
https://doi.org/10.1007/978-981-96-9782-3_13

display (such as an OLED or LCD display) directly in front of the eye with suitable optics to create an image that appears to be emitted from in front of the user. Such displays have resulted in commercial systems from Apple, Meta, etc. By showing different views to the left and right eye a stereoscopic 3D effect can be created. Furthermore, VR can offer excellent color quality, field of view and contrast. However such headsets are cumbersome, isolate the viewer from the real world, and are thus normally limited to situations where the viewer is seated, for example, for computer gaming, or for 3D visualization. In contrast augmented reality (AR) places artificial scene light into the visual field of the user. This is more difficult to achieve as an optical approach is needed to combine the artificial scene and natural scene light. AR glasses need to be close to regular spectacles in terms of size, weight, visual appearance, and this makes issues such as brightness, transparency, battery life particularly critical. However, AR is an approach that can be used in ordinary life, while standing, walking, exercising etc. Such freedom lends itself to myriad applications including receiving messages and email, subtitling real-life conversation, automatic speech to text translation, training, medical data, etc as well as entertainment, gaming, etc.

Linear vs nonlinear combining. The current leading approach for creating AR displays is to make use of waveguides and output coupling gratings. This approach has been developed widely by companies such as Meta, SNAP, MagicLeap, Digilens, and others. Other schemes involve projection off a meta-surface or diffraction grating of free-space beams, while others involve incorporating reflective elements within the material of the lens. All of these approaches work by scattering of light, which is a linear optical property. The challenge of all these approaches is that a material cannot be simultaneously transparent and scattering, or in other words it cannot scatter artificial light to the eye without also scattering some natural scene light that was travelling towards the eye away from the eye. These linear optical scattering effects cause firstly eye-glow which is light emitted from the display into the external scene. Eye-glow is problematic because it can be seen externally, leads to lack of privacy, which in defense and security can be dangerous, and is distracting to people in a conversation. Secondly, linear optics scattering causes rainbow which is scattering of natural scene light into different angular positions seen by the wearer of the AR display. In the case of white light sources and diffractive approaches this creates rainbow visual effects – hence the name. Thirdly, the scattering effect of linear optics causes the display to be tinted or less than 100% transparent, this is visually unappealing to consumers and also limits optical (and hence electrical) efficiency. While approaches such as volume Bragg gratings and high index materials can help with all these effects the fundamental challenge remains and is due to the linear optical approach. As we will show – using nonlinear optical wave-mixing can provide an alternative route to placing artificial scene light into the natural light path, and thus provides a potential route to AR glasses that are 100% transparent and with zero eye-glow or rainbow. Additionally, the nonlinear approach also allows features such as wavefront control to be implemented.

2 Nonlinear optical display

2.1 Experimental implementation

Our display works by placing a nonlinear material within an optical set-up that combines two infrared laser beams. In our experiments the 1030nm beams are combined in LiNbO3 in a non-colinear geometry to create a green virtual image that is emitted at normal incidence from the display element. One beam is encoded with intensity information using a spatial light modulator (SLM) which is driven by a video signal, and the second beam mixes with the first to create light at the second harmonic wavelength (green) which is viewed by the eye. The input laser is a diode pumped solid state laser and the beam is split into two paths. Lens are used to relay the light from the into the nonlinear crystal which is placed in the Fourier plane of the system. When it mixes with the second beam it creates an image localized at infinity. The image can be viewed by eye, or by using a camera (as in the image shown below). If the radius of curvature of the second beam is altered (for example using electrically controlled lenses) the resulting virtual image can be altered from being at infinity to a finite distance. We have demonstrated image localization from infinity down to 30cm which corresponds to the near-point of normal vision.

We have also demonstrated Red, Green and Blue (RGB) within the same nonlinear optical element by using three infra-red lasers. Nd doped diode pumped solid state lasers are ideal for this using different lasing transitions. Our approach lends itself to a sequential color scheme for creating color images, and furthermore, using the electrically controlled lens we can also create place images with different virtual depth. We believe that with appropriate engineering, and using sequential color and depth frames, it will be possible to create realistic full-color displays which will have proper depth cues (accommodation eye focusing) and should also help to eliminate nausea created by the vergence / accommodation conflict.

Our demo device uses LiNbO3 which is essentially 100% transparent in the visible, although other materials such as other inorganic or organic NLO materials could also be used. Indeed, such NLO materials are necessarily almost perfectly transparent or they would have compromised use in conventional nonlinear optics where Watt level lasers (and much higher) are routinely used, and any absorption would result in heating and material damage. Fortunately for the display application the output powers are in the micro-Watt regime, and so damage will not be an issue. Of course, by using DPSS lasers which give sub-nanosecond pulses we can achieve high peak power (kW) to drive the nonlinear optical process, while operating at cw milliwatt levels which ensure both eye-safety and also low electrical consumption

Fig. 1. The image shows the University of Southampton logo created by our novel display. The image was taken during the Photonics West Conference Jan 2025 and is overlaid over a view of San Francisco above the Moscone Center.

3 Conclusion

We have demonstrated a new type of close-to-eye display based on nonlinear optics. In lab prototypes we have demonstrated eye-viewable displays capable of RGB operation. The displays are bright, essentially 100% transparent, do not have any unwanted diffractive artifacts (such as eye-glow or rainbow) and, as they are based on wavemixing are inherently directional and thus optically and electrically efficient.

By using Fourier optics approaches and wavefront control it is possible to create virtual images that are localized at any desired distance between infinity (far-field) and the near-point of the human eye. We believe our patented approach [2] provides a route to creating AR glasses that will eventually look exactly like normal spectacles. They will be as transparent as regular spectacles, will 'power-off' into fully transparent mode, and will offer the performance in terms of resolution. SWAPC (size-weight-power-cost) that will enable convincing AR glasses to become a consumer product.

We would like to acknowledge the support of EPSRC, DASA, DSTL, University of Southampton and Innovate UK for project funding.

References

1. Augmented reality and virtual reality displays: emerging technologies and future perspectives Jianghao Xiong, En-Lin Hsiang, Ziqian He, Tao Zhan and Shin-Tson Wu - Light: Science & Applications (2021)10:216. https://doi.org/10.1038/s41377-021-00658-8
2. WO2023241985 - DISPLAY APPARATUS

Part III
Joint Session "Light Sources and Components" and "Laser Applications for Moving Platforms" LDC3

Measurement of properties of light generated in laser lighting

Ole Bjarlin Jensen[1[0000-0002-4295-5951]], Jian Xu[2[0000-0002-4880-6770]], Dennis Dan Corell[1[0000-0002-1929-2344]] and Carsten Dam-Hansen[1[0000-0001-7518-4025]]

[1] Diode Lasers and LED Systems Group Department of Electrical and Photonics Engineering, Technical University of Denmark, 4000 Roskilde, Denmark
[2] Lab of New Energy Materials and Devices, School of Physics and Electronic Information, Henan Polytechnic University, Jiaozuo 454000, China.
ojen@dtu.dk

Abstract. Light generated by laser induced luminescence is promising for many applications requiring high luminance. Some properties of the generated light are similar to well-known light sources while other properties are significantly different. These properties may require special measurement methods in order to properly extract the correct information about the light. In this paper, we will introduce some of the measurement methods applicable to characterize light generated by laser induced luminescence.

Keywords: Laser lighting, phosphor materials, light characterization.

1 Introduction

Laser lighting is an emerging lighting technology, which shows great promises for enabling high luminance light sources [1]. The underlying mechanism used is photoluminescence similar to light emitting diodes (LEDs). In a white LED, a blue LED is used to excite carriers in luminescent materials and when relaxing to the ground state a lower energy photon is emitted. The mixing of the blue light and the longer wavelength luminescent light generates white light. In laser lighting, the process is similar except that a blue laser is used to excite the luminescent material. This small difference changes the spectral content of the white light as the laser has a narrow spectrum. Furthermore, the laser is a coherent light source capable of causing interference, which induces speckle in the generated white light. The properties of the luminescent material also have a large impact on the generated light. The spectral content and properties will change with different luminescent materials and the appearance of the light is affected by the structure of the luminescent material [2,3]. The surface and microstructure of the luminescent material has a significant impact on the achievable luminance as it may help in confinement of the emission spot [4]. All these properties of the generated light require different measurement methods, which to some extend are different than for LEDs.

S. Kurimura and T. Yagi (eds.), *Proceedings of Laser Display and Lighting Conference 2025*, Springer Proceedings in Physics 429,
https://doi.org/10.1007/978-981-96-9782-3_14

In this paper, we will introduce some of these measurement methods with examples on measurement on various luminescent materials. The focus is on measurement of the luminescence, while measurements of material properties are not covered.

2 Characterization of phosphor materials for laser lighting

Phosphors are the luminescent material of choice in most laser lighting applications. Therefore, characterization of the luminescent properties of phosphor materials is mandatory when developing new light sources based on phosphors. Here, we introduce some of the characterization methods available for this.

The luminescent properties that we will focus on in this paper are:

- Luminous flux and efficiency. These properties are important to achieve the highest efficiency from the light source. The luminous flux emitted by a phosphor material is most accurately measured using an integrating sphere spectroradiometer being able to collect all the emitted light. From this setup, the spectral radiant flux $\Phi_{e,\lambda}(\lambda)$ is measured and from this, the luminous flux Φ_V can be calculated by integration using the photopic eye sensitivity function $V(\lambda)$.

$$\Phi_V = 683\frac{lm}{W} \cdot \int_{VIS} V(\lambda)\Phi_{e,\lambda}(\lambda)d\lambda \quad (1)$$

- Luminescent spot size. The luminescent spot size on the phosphor material determines the luminous exitance (emitted flux pr surface area), which in terms determines the achievable luminous intensity and luminance of a light source. Therefore, the ability of a phosphor mate-rial to confine the emission spot size is important in order to enable high luminance. Scattering at the surface and inside of the phosphor material will enlarge the emission spot and also the absorption properties and thickness of the phosphor will affect the emission spot size [5].
- Luminous spot illuminance and chromaticity. Depending on the homogeneity and particle size in the phosphor material, the white light emitted from the phosphor will be more or less homogeneous. Researchers have described the emitted light as having e.g. a yellow ring around the central spot [2,6,7]. This may be caused by the yellowish luminescent emission spot being larger than the blue light re-emitted from the phosphor. The homogeneity of the emission may be investigated in detail by collecting the emitted light with a suitable lens and investigating the spot using a luminance and color camera.
- Speckle in laser lighting. Lasers are coherent light sources and as such capable of creating interference patterns. Speckle is an interference pattern created by interference between a large number of wavelets scattered off a surface. Speckle is an unwanted effect in laser lighting as it may cause disturbances when viewing. The speckle contrast is generally used as a measure for the amount of speckle in light [8]. There are different ways to measure the amount of speckle in light generated by laser lighting. Here, we consider objective speckles, which are observed on a plane illuminated by the white light.

Various examples of measurement results on different phosphors are given in Fig. 1. The measurement methods and results will be discussed in detail in the presentation.

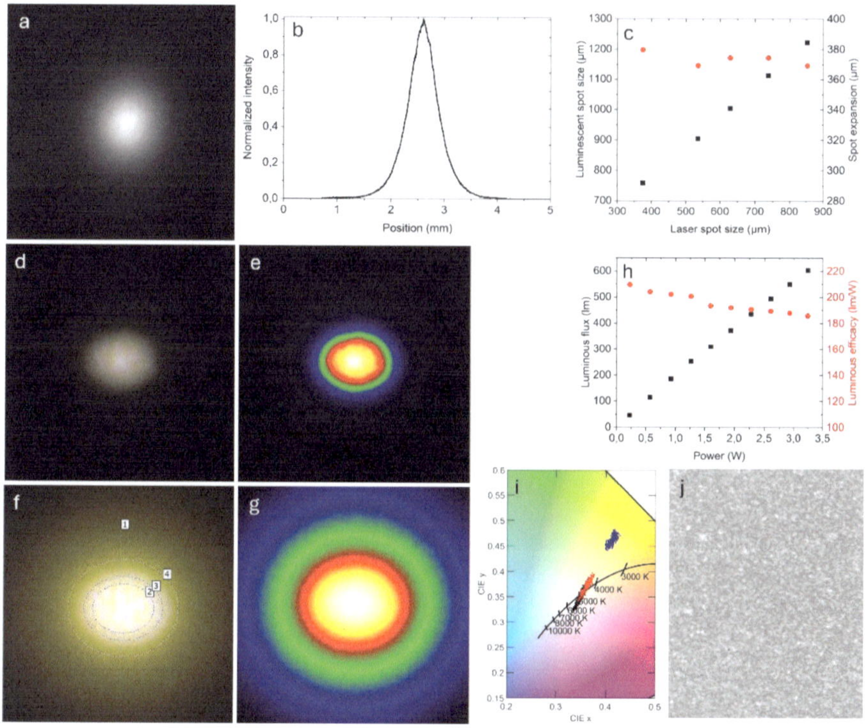

Fig. 1. Example results for measurement of luminous spot image (a), cross section (b) and spot size and expansion (c) for a composite ceramic phosphor; high-dynamic range color (d, f) and luminance images (e, g) and chromaticity coordinates in the CIE 1931 chromaticity diagram (i) for a phosphor in glass film (PiGF); luminous flux and efficacy for a composite ceramic phosphor (h); speckled image of the white light emitted from a PiGF (j).

References

1. J. J. Wierer, J. Y. Tsao, and D. S. Sizov, "Comparison between blue lasers and light-emitting diodes for future solid-state lighting," Laser Photon. Rev. **7**, 963–993 (2013).
2. S. Li, L. Wang, N. Hirosaki, and R. J. Xie, "Color Conversion Materials for High-Brightness Laser-Driven Solid-State Lighting," Laser Photonics Rev. **1800173**, 1–29 (2018).
3. O. B. Jensen, X. Chen, J. Xu, D. D. Corell, and C. Dam-Hansen, "Luminance and chromaticity characteristics of different phosphor types in laser lighting," JPhys Photonics **6**, 015005 (2024).
4. J. Xu, L. Wang, W. Gu, Z. Jiang, X. Chen, B. Hu, B. Du, H. Ji, C. Dam-Hansen, and O. B. Jensen, "Emitting area limitation via scattering control in phosphor film realizing high-luminance laser lighting," J. Eur. Ceram. Soc. **42**, 608–615 (2022).

5. A. Krasnoshchoka, A. K. Hansen, A. Thorseth, D. Marti, J. Xu, P. M. Petersen, and O. B. Jensen, "Phosphor material dependent spot size limitations in laser lighting," Opt. Express **28**, 5758–5767 (2020).
6. Y. Ma and X. Luo, "Small-divergent-angle uniform illumination with enhanced luminance of transmissive phosphor-converted white laser diode by secondary optics design," Opt. Lasers Eng. **122**, 14–22 (2019).
7. S. Feng, Y. Guo, M. Allix, S. Li, R. J. Xie, J. Fu, C. Genevois, E. Véron, H. Wang, Y. Yang, H. Qin, and J. Li, "Biphasic Lu3MgAl3SiO12-based transparent ceramics for uniform laser-diode-driven white lighting," Cell Reports Phys. Sci. **3**, 101044 (2022).
8. J. W. Goodman, "Some fundamental properties of speckle," J. Opt. Soc. Am. **66**, 1145–1150 (1976).

Asynchronous Optical Sampling for High-Resolution LiDAR

Hollie Wright[1], Alexander J. M. Nelmes[1] and Derryck T. Reid[1]

[1] Institute of Photonics and Quantum Sciences, School of Engineering and Physical Sciences, Heriot-Watt University, Edinburgh, EH14 4AS, UK.
h.wright@hw.ac.uk

Abstract. With asynchronous optical sampling, we overcome the spatial resolution of time-of-flight LiDAR by up to three orders of magnitude. Here we present recent demonstrations of multi-target ranging, non-optical target ranging, and one-dimensional surface scanning.

Keywords: LiDAR, time of flight, absolute distance measurement, dual-comb ranging, asynchronous optical sampling, frequency combs.

1 Introduction

Time-of-flight (ToF) LiDAR (light detecting and ranging) is the dominant optical metrology technique for metre-scale use-cases. Emitted light reflects off objects and returns to the detector; the elapsed time is recorded to allow calculation of the optical distance to the object. The technique has been broadly applied in airborne, terrestrial and underwater applications including forestry, surveyance, and autonomous vehicles. However, state-of-the-art ToF LiDAR systems have a depth resolution typically limited to 0.1-1 cm [1], so there is a need for higher resolution LiDAR for certain applications. In response to this, we are developing Two-Photon Dual-Comb LiDAR (2phDCL) as the basis for the next generation of LiDAR technologies. Here, we introduce the 2phDCL concept. In Section 2, we present demonstrations of ranging to multiple dynamic targets; and in Section 3, we demonstrate ranging to non-optical, (specifically, brass) targets. In Section 4 we present an outlook towards future industrial applications of the technique; Section 5 concludes the discussion.

1.1 Introduction to Two-Photon Dual-Comb LiDAR

Our technique is a variation of dual-comb ranging (DCR); A typical DCR experiment is visualized in Fig. 1(a). Pulses from the probe comb enter the measurement arm via the first beam splitter (BS) cube, ands generate reflections off reference and target optics. The reflections are combined with pulses from the LO comb at the second BS cube so that both combs are detected together by a photodiode. If the repetition rates of the probe and LO pulses are sufficiently similar (but not identical) the pulses effectively walk through each other, creating cross-correlations when they overlap.

S. Kurimura and T. Yagi (eds.), *Proceedings of Laser Display and Lighting Conference 2025*, Springer Proceedings in Physics 429,
https://doi.org/10.1007/978-981-96-9782-3_15

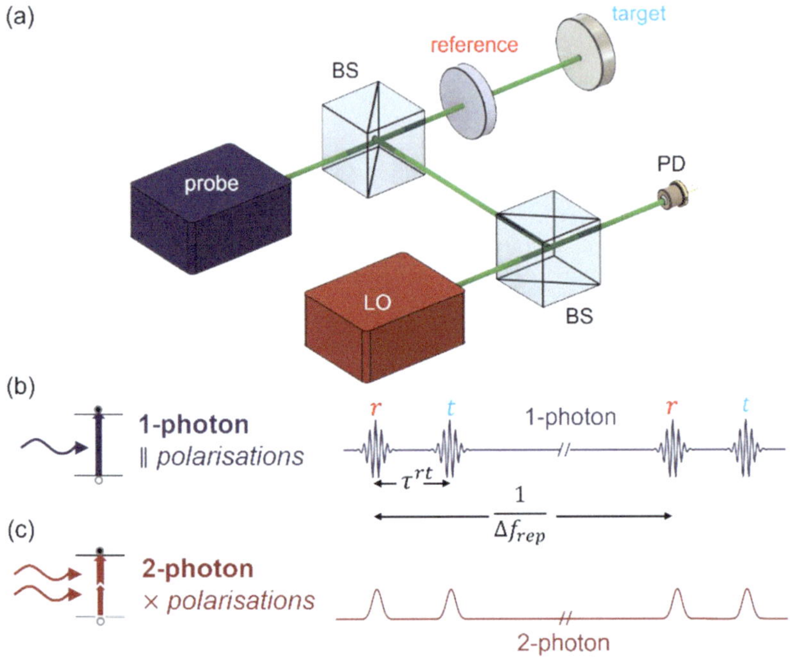

Fig. 1. (a) Basic DCR set-up. BS: beam splitter; PD: photodiode. (b) Conventional one-photon detection of interferograms. (c) Two-photon detection of intensity cross-correlations as implemented in 2phDCL.

When conventional one-photon detection is used, first-order interferometric cross-correlations ("interferograms") are generated when the probe and LO pulses temporally overlap. An electronic low-pass filter (LPF) with a cut-off frequency at roughly half the repetition rate, f_{rep} is applied to remove the probe and LO pulse trains and leave only the interferograms, as illustrated in Fig. 1(b) [2]. The signal consists of pairs of interferograms, defining the reference (r) and target (t) positions, repeating at a rate equal to the difference in the combs' repetition rates, Δf_{rep}. The timing gap between interferogram pairs, τ^{rt} quantifies the time-of-flight between the reference and target optics, allowing the optical distance, d to be calculated using Eqn. 1.

$$d = \frac{v_g \tau^{rt}}{2} \frac{\Delta f_{rep}}{f_{rep}} \tag{1}$$

In which v_g is the group velocity of the pulses. In most cases it is acceptable to approximate v_g as the speed of light constant, c. Depending on the stability of the probe and LO sources, the carrier frequency of the interferograms can vary rapidly in time, which often makes it necessary to digitize the full optical bandwidth at speeds typically >100 MSa/s. Following digitization, data analysis typically involves applying Hilbert transforms and fitting gaussians to each interferogram to determine the centre point for high-resolution determination of τ^{rt}. The result is a powerful high-precision ranging technique, but the large data burden and the complex data processing have largely limited DCR to lab-based investigations which do not provide real-time ranging information.

To develop DCR towards industrial applications we developed 2phDCL, a non-interferometric variation of DCR which detects second-order intensity cross-correlations rather than the conventional interferograms, as illustrated in Fig. 1(c) [3]. These second-order signals are detected via two-photon absorption to provide a strong nonlinear signal. We ensure non-interferometric cross-correlations by detecting the probe and LO pulses with orthogonal polarisations so they cannot optically interfere. The carrier-free nature of the cross-correlations means they can be electronically conditioned to take the form of CMOS-compatible impulses, which can be detected as a rising edge by a microcontroller. We programmed a Teensyduino 4.0 to count its internal clock cycles and save the count whenever a rising edge was detected, thus allowing us to quantify the time delay between subsequent cross-correlations as two repeating integers: τ^{rt}, the delay between the reference and target cross-correlations; and τ^{tr}, the delay between the target and the next reference cross-correlation. The raw data takes the form of two repeating integers printed to the serial port by the microcontroller. We calculate the distance, d using Eqn. 2.

$$d = \frac{v_g}{2f_{rep}} \frac{\tau^{rt}}{\tau^{rt}+\tau^{tr}} \tag{2}$$

The reduction in the complexity of the data analysis, plus the reduction in the data burden, significantly simplifies the DCR technique. This, coupled with our compact, eye-safe comb sources, makes our system particularly suitable for field deployment.

2 Simultaneous ranging to multiple moving targets

Our simple data collection / analysis is easily extended for simultaneous ranging to multiple targets using Eqn. 3.

$$d = \frac{c}{2f_{rep}} \frac{\sum_{i=1}^{n} \tau_i}{\sum_{i=1}^{n+1} \tau_i}, \tag{3}$$

where n is the number of targets. Our experimental configuration is shown in Fig. 2. Pulses leave the probe comb with elliptical polarisation so a quarter- and half-waveplate are introduced to maximise transmission through the first polarising beam splitter (PBS). A reference reflection is provided by the BS plate; the rest of the probe passes through a diffractive optical element (DOE) and is split into the zeroth and $\pm 1^{st}$ order diffracted beams. The now-three probe beams each sample a target retroreflector and the three reflections are recombined at the DOE. The reference and three target reflections are reflected at both PBSs since they were translated to perpendicular polarisation by the double pass through the quarter-waveplate. The LO pulses also leave the comb source with elliptical polarisation and are translated to parallel polarisation by passing through quarter- and half-waveplates. The probe reflections and LO pulses are therefore combined with orthogonal polarisations before being focused onto a photodetector. We use a laser diode (LD) as the detector as we have found LDs offer a stronger nonlinear response than PDs [5,6].

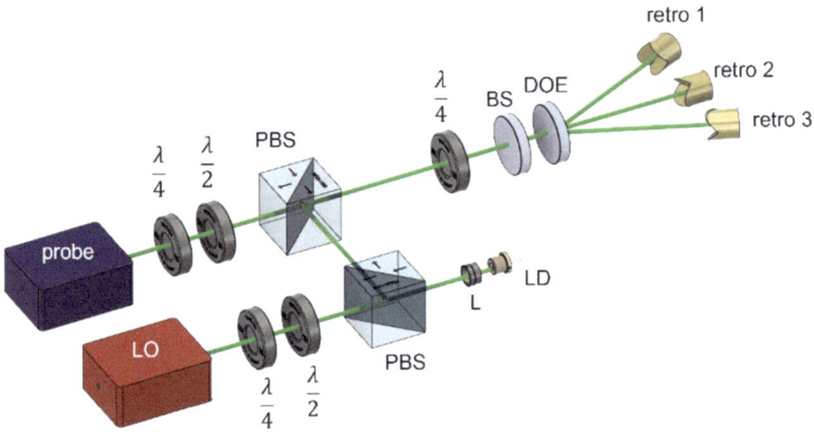

Fig. 2. Experimental set-up for simultaneous ranging to multiple targets. PBS: polarising beam splitter; DOE: diffractive optical element; LD: laser diode.

To create moving targets, we mounted each of the retroreflectors on simple loudspeakers driven by low frequencies from signal generators. Figure 3(a) shows a photograph of one such retroreflector mounted on the front of a speaker, which is held in a 3D printed mount. Figures 3(b)-3(d) show the simultaneously collected ranging data for the three targets, driven by sinusoidal signals at 1, 2 and 3 Hz. Each plot is displayed with a consistent distance axis spanning 1.5 cm. The blue data show single-shot measurements collected approximately every 1 ms. The red data show the results with averaging for 20 ms.

The time-averaged data in Figs. 3(b)-3(d) clearly show sinusoidal motion at the expected frequencies. Comparing the single-shot data to the time-averaged data shows minor variations around the expected sinusoidal motion. This was likely caused by vibration of the retroreflectors during the motion. After offsetting the time-averaged motion, the standard deviations of the single-shot measurements were 78 μm, 83 μm and 84 μm respectively. A potential application for this dynamic target sensing is proposed in Section 4.1.

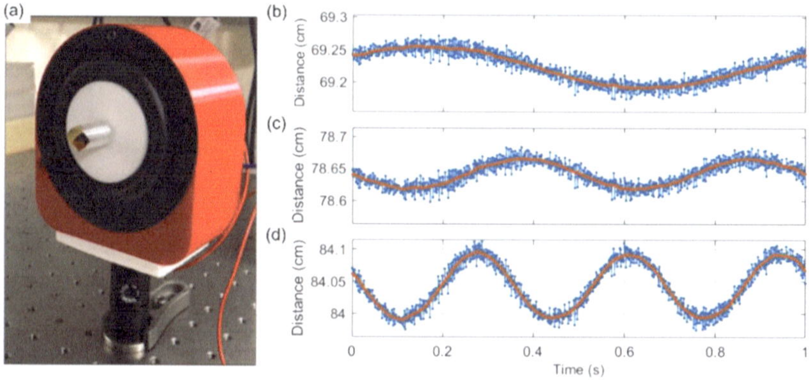

Fig. 3. (a) Retroreflector target mounted onto a loudspeaker to create a dynamic target. (b)-(d) Data showing simultaneous ranging to three retroreflectors oscillating sinusoidally at frequencies: (b) 1 Hz; (c) 2 Hz; (d) 3 Hz. All subplots have the same distance span of 1.5 cm. The blue data show single shot ranging; the red data show ranging with 20 ms of averaging.

3 Ranging to noncooperative targets

3.1 Single point ranging

Some of the intended applications for our technique require ranging to non-cooperative targets such as metals. The rough surfaces of non-cooperative targets provide very weak optical reflections, making measurement difficult. We incorporated amplified detection into our acquisition system to recover high-precision ranging. Figure 4 shows the fiberised version of our DCR system, which includes an Erbium-doped fibre amplifier (EDFA) to amplify weak target reflections. Probe pulses pass through a circulator and are emitted into free-space at an FC/PC connector. An aspheric lens is used to approximately collimate the beam. A BS plate provides a reference reflection, and the remainder of the beam reflects off the target. The reflections are coupled back into fibre at the FC/PC connector and pass again through the circulator to be amplified in the EDFA. The amplified probe reflections are combined with the LO pulses in a fibre combiner. As in the free-space implementation, the probe and LO are combined with orthogonal polarisations to eliminate optical interference. An electronic LPF removes the probe and LO pulse trains from the signal, leaving only the intensity cross-correlations to be electronically amplified. The weak returns from the metal target lead to an amplitude difference in the reference and target cross-correlations; a constant fraction discriminator (CFD) takes the uneven cross-correlations and outputs CMOS-compatible impulses which are read by our Teensy 4.0 microcontroller.

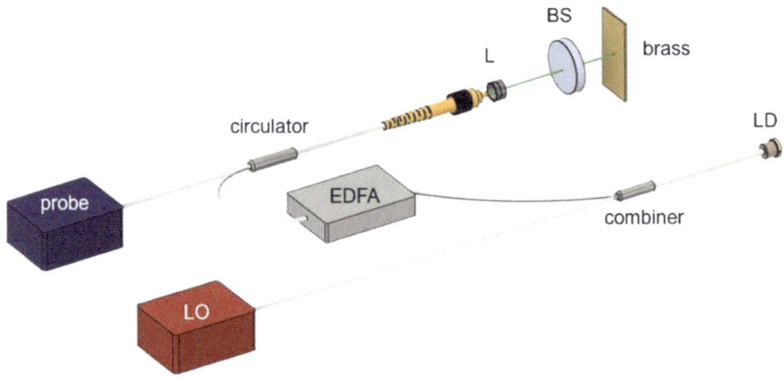

Fig. 4. Fiberised system with EDFA for amplifying weak optical returns.

Ranging data were collected for distances between 2 cm and 1 m. For distances beyond 50 cm, the probe beam was focused onto the brass surface to maximise optical return. The standard deviation of each data set quantified the measurement precision. Figure 5(a) shows the standard deviation as a function of measured distance for the brass target, plus a silver mirror for reference. The precision for both silver and brass was below 20 μm for distances under 10 cm. The precision worsened to ≥100 μm when the targets were moved to 1 m away. The plot shows a near-linear relationship between precision and distance. We believe this is caused by a reduced coupling efficiency when focusing the reflections back into the FC/PC connector. Figure 5(b) shows the Allan deviations for the silver and brass targets, which quantify the precision to be gained by increasing measurement averaging time. Both data sets were collected at a distance of 20 cm. Again, both targets show a similar relationship. For single-shot measurements without averaging, the measurement precisions were 37 μm and 40 μm for the brass and silver, respectively. After 500 ms of averaging, the precisions improved to 690 nm and 763 nm, respectively. The similarity of the brass and silver data presented in Fig. 5(a)-5(b) confirms that amplified detection recovers high-precision ranging in noncooperative targets.

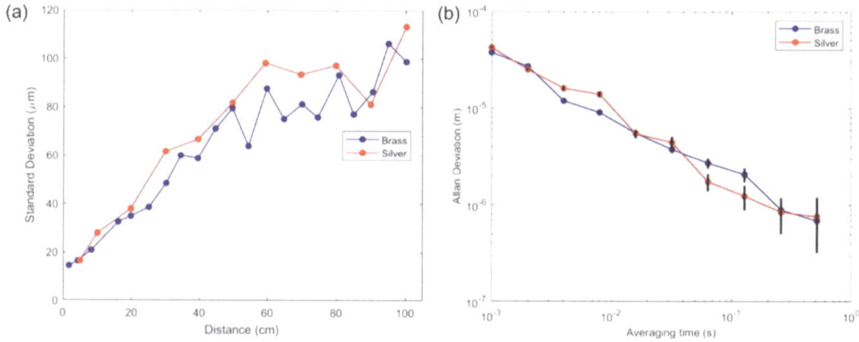

Fig. 5. (a) Ranging precision as a function of measured distance for brass and silver. (b) Allan deviations for brass and silver data sets.

3.2 Line-scan ranging

To create a structured target, we milled a stepped artefact out of a small piece of brass. Figure 6 shows a photograph of the artefact. The piece was not precision-milled so there are visible variations in the stepped structure, including sloping downwards/upwards at several step edges.

Fig. 6. Brass steps milled to approximate step heights of 1mm.

The artefact was mounted onto a motorised translation stage and ranging data were collected as the probe beam translated across the stepped surface. Figure 7 shows the distance measured to the brass surface, as a function of the translation stage position, creating a one-dimensional surface profile of the artefact. The plot is laid over a side-profile image of the artefact, captured by placing the object on the bed of a computer scanner. The plot recreates the side profile except for one erroneous data point at x = -11 cm. LiDAR data were collected by scanning across the approximate centre of the steps, rather than the side-edge shown in the scanned image so the data are not expected to perfectly align with the scanned side-profile.

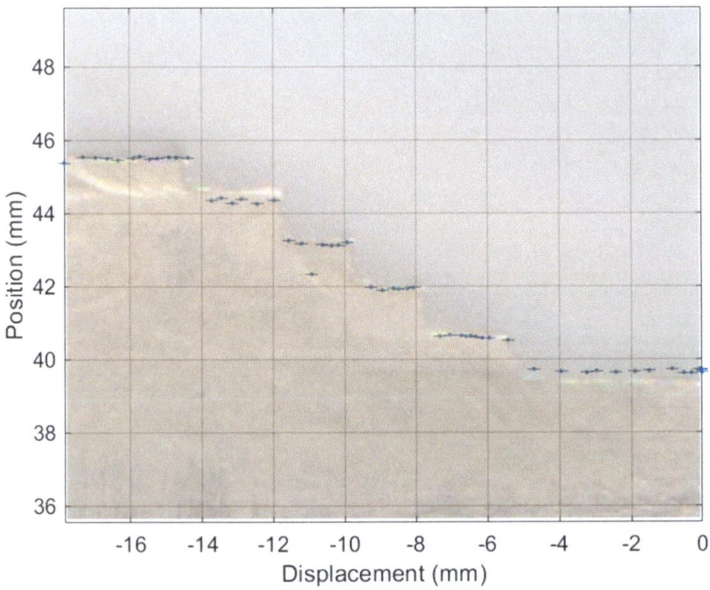

Fig. 7. LiDAR data recreate the surface profile. The plot is laid over a computer-scanned image of the artefact.

Figure 7 provides a proof-of-concept demonstration of one-dimensional surface scanning of a metal artefact. In future work, we will extend the measurement to two-dimensional surface scanning.

4 Conclusion and Outlook

We have presented recent results demonstrating the use of 2phDCL as a high-resolution LiDAR technique: for simultaneous ranging to multiple optical targets; for ranging to brass; and for one-dimensional surface scanning of a brass artefact. These demonstrations lay the foundation for future applications in manufacturing and artefact characterisation, upcoming work will share results for other noncooperative targets. The demonstrations were performed with 1.5 μm modelocked lasers, as these are most suitable for our immediate applications. However, it should be possible to perform our 2phDCL technique at visible wavelengths, provided a LD with an appropriate bandgap energy can be obtained.

The demonstration of simultaneous ranging to multiple dynamic targets provided in Section 2 lays the groundwork for absolute positioning of moving objects. One specific application for this is tracking the position of a manufacturing machine's end-effector. The motion of manufacturing machines is predetermined by internal controls which typically have up to six degrees of freedom. These controls are vulnerable to a plethora of error sources, such as vibration, backlash, and environmental changes. If positional errors are so significant that the manufactured item falls outwith design tolerances, the item must be scrapped. This can lead to considerable waste, particularly in high-value manufacturing, which limits the sustainability of these manufacturing processes. We anticipate our system could provide external position verification for a machine's end-effector, which could be used in a closed-loop system to compensate for manufacturing missteps in real time.

References

1. Royo, S., & Ballesta-Garcia, M.: An overview of lidar imaging systems for autonomous vehicles. Applied sciences, 9(19), 4093 (2019).
2. Coddington, I., Swann, W. C., Nenadovic, L., & Newbury, N. R.: Rapid and precise absolute distance measurements at long range. Nature photonics, 3(6), 351-356 (2009).
3. Wright, H., Sun, J., McKendrick, D., Weston, N., & Reid, D. T.: Two-photon dual-comb LiDAR. *Optics Express*, *29*(23), 37037-37047 (2021).
4. Wright, H., Nelmes, A. J., Weston, N. J., & Reid, D. T.: Multi-target two-photon dual-comb LiDAR. Optics Express, 31(14), 22497-22506 (2023).
5. Reid, D. T., Sibbett, W., Dudley, J. M., Barry, L. P., Thomsen, B., & Harvey, J. D.: Applied Optics *37*, 8142 (1998).
6. Sterczewski, Ł. A., & Sotor, J.: Two-photon imaging of soliton dynamics. Nature Communications, 14(1), 3339 (2023).

Direct emitting diode laser modules below 630 nm for display applications

Katrin Paschke[1], Felix Mauerhoff[1], André Maaßdorf[1], David Feise[1], Oktay Senel[1] and Gunnar Blume[1]

[1] Ferdinand-Braun-Institut (FBH), Gustav-Kirchhoff-Str. 4, 12489 Berlin, Germany
katrin.paschke@fbh-berlin.de

Abstract. This paper aims to develop direct emitting diode laser sources for display applications like automotive backlight displays. The focus is placed on emission wavelength below 630 nm due to an increase in luminance and foremost due to the visibility for red-green color-blind people. We use a macroscopic setup for gain chip evaluation and finally build a miniaturized fiber coupled laser module emitting 10 mW at 619 nm.

Keywords: automotive displays, laser displays, color-blind, diode laser module, 619 nm

1 Introduction

Automotive backlight systems are of high interest for industrial applications. These systems require high luminance light sources in the red, green and blue spectral region. With decreasing wavelengths towards orange, the eye sensitivity increases, which is one reason for the interest of shifting red emitting diode laser towards orange emission [1]. The other reason is that red-green color-blind people benefit from displays operating below 625 nm. I think that laser diodes emitting at 619 nm are very suitable for such automotive application. When LED backlighting was first installed in cars, the center wavelength was about 640 nm, but the emission wavelength was too long for red-green color-blind people to properly see the displayed information. Therefore, LED backlight manufacturers adjusted the center wavelength towards 620 nm. The development of short wavelength red laser diodes has its limitations in the carrier confinement in the active region, which has been the a research focus for several decades now [2], [3], [4].
In this paper we show the development of direct emitting diode lasers below 625 nm by using temperature tuning of the gain peak. The paper is structured as follows: first, we introduce the used gain chip, followed by the experimental setup. We use a table-top setup for general feasibility studies. Based on these findings a miniaturized laser module emitting at 619 nm was developed. At the end of the paper, the module properties are presented and all results are summarized.

S. Kurimura and T. Yagi (eds.), *Proceedings of Laser Display and Lighting Conference 2025*, Springer Proceedings in Physics 429,
https://doi.org/10.1007/978-981-96-9782-3_16

2 Gain chip

In this section, the gain chip and its vertical and lateral structure are introduced.

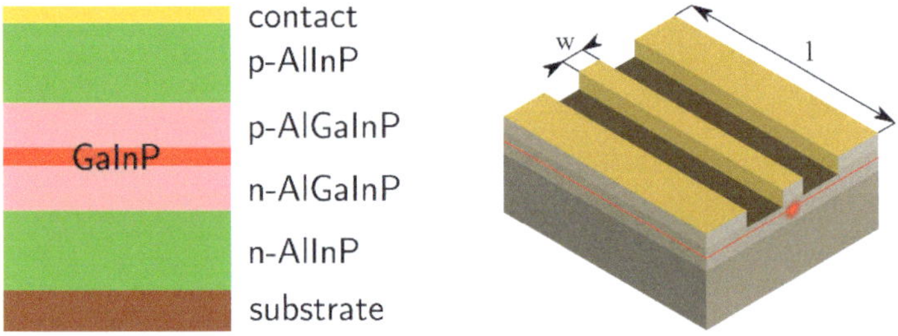

Fig. 1 Epitaxial structure for room temperature emission below 630 nm (left) and drawing of ridge waveguide emitter with a ridge width *w* and resonator length *l* (right)

The vertical structure is schematically depicted in Fig. 1 (left). We utilize a 960 nm wide symmetric $Al_{0.36}Ga_{0.16}In_{0.48}P$ waveguide, that is embedded in an $Al_{0.52}In_{0.48}P$ cladding with a total thickness of 1500 nm. The active zone is composed of an 8 nm wide $Ga_{0.64}In_{0.36}P$ single quantum well. This structure is optimized for room-temperature operation at 626 nm, as reported in [5]. The lateral structure is shown in **Fig. 1** (right). A 7.5 µm wide ridge and 2 mm long resonator are utilized. The ridge is fabricated by RIE. The etch depth is chosen to change in the effective refractive index of ~5 x 10^{-3}. This promotes spatial single mode operation, as shown with ridge waveguide lasers of the same vertical and lateral structure [6]. The chip presented here is antireflection coated on the front facet.

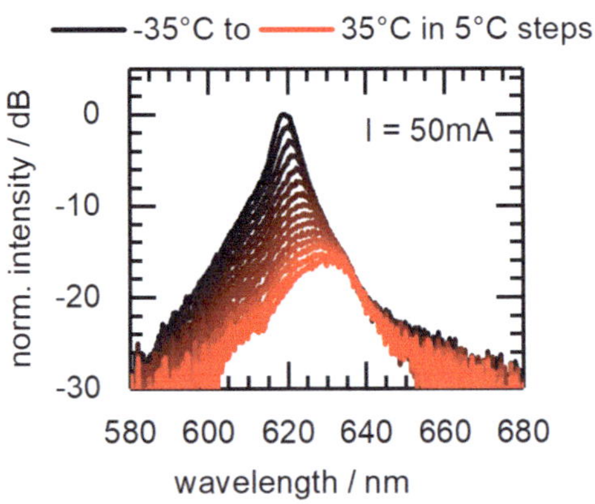

Fig. 2 Temperature dependent amplified spontaneous emission spectra at heat sink temperatures from 35 °C to -35 °C in 5 °C steps.

In order to achieve laser emission below 626 nm the gain chip is cooled. **Fig. 2** shows the shift in the amplified spontaneous emission spectra with respect to heat sink temperature. The gain peak shifts to shorter wavelengths with a rate of ~ 0.15 nm / °C, reaching 619 nm at ca. -35 °C.

3 Experimental setups

We use two different setups to verify the feasibility of building laser modules emitting below 630 nm. First, a tabletop experiment of an externally wavelength stabilized diode laser (ECDL) was created. The Littman-Metcalf technique is employed using a reflective grating with 2400 lines per mm. Second, a miniaturized laser module was built based on a fiber-Bragg-grating with a center wavelength of 619.1 nm and, a FWHM of 100 pm and, a peak reflectivity of ~ 10 %. The grating is written into a polarization maintaining single mode fiber. The module is based on a 14-pin standard butterfly module. The optical setup consists of two acylindrical lenses for fast and slow axis collimation of the gain chip. Another set of two acylindrical lenses is used to couple the light into the fiber. The focal lengths are selected to optimize the coupling efficiency of the strongly asymmetric beam.

4 Measurement results

In this section we want to give a brief overview of the emission properties of the tabletop and the laser module experiments. The tabletop setup allows for wavelength tuning down to 619 nm as shown in **Fig. 3**.

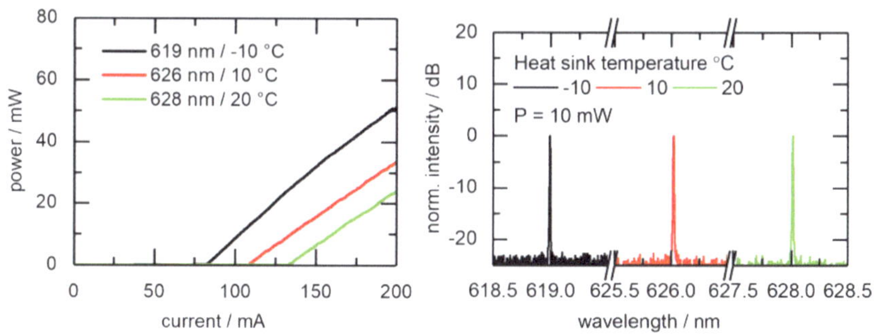

Fig. 3 Power-current characteristic of macroscopic ECDL setup at different heatsink temperatures and wavelengths (left) and respective emission spectra at a constant optical power of 10 mW (right)

Continuous wavelength tuning while adapting the heat sink temperature from 20 °C to -10 °C, allows tuning the emission wavelength from 628 nm to 619 nm. The left plot shows the power current characteristics at constant temperatures and wavelengths. The right plot shows normalized emission spectra at a constant heat sink temperature and

optical output power of 10 mW. Here, a single longitudinal mode emission is observed at the wavelengths of 628 nm, 626 nm and 619 nm and heatsink temperatures of 20 °C, 10 °C and -10 °C, respectively. We measured respective peak output powers of more than 20 mW, 30 mW and 50 mW at an injection current of 200 mA.
The miniaturized laser module targets 619 nm and respective measurement results are shown in **Fig. 4**. The left plot shows the power-current characteristics for various internal temperatures from 0 °C to -25 °C in 5 °C steps, while the external temperature was held constant at 20°C. The module emits 10 mW from the fiber at an internal temperature of -10 °C at an injection current of 150 mA. The right graph shows the single longitudinal emission at an inlay temperature of -10 °C and an injection current of 135 mA and an output power of ca. 6 mW. A side mode suppression of more than 30 dB can be seen.

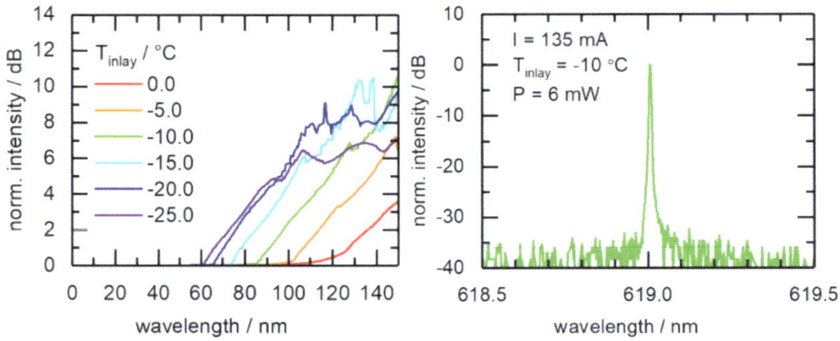

Fig. 4 Characterization results of the miniaturized laser module emitting at 619 nm (left) power current characteristic at different inlay temperatures (right) emission spectrum at an inlay temperature of -10 °C and injection current of 135 mA

5 Summary

We have shown continuous wavelength tunability by adapting the heat sink temperature of the gain chip from above 628 nm down to 619 nm, reaching output powers at 619 nm of about 50 mW in an ECDL setup. We successfully designed and built a miniaturized laser module emitting more than 10 mW of output power out of a polarization maintaining single mode fiber at 619 nm. The gain chip is housed in a 14-pin butterfly module using a polarization maintaining single mode fiber-Bragg-grating for wavelength stabilization.
This shows the feasibility of employing laser diodes in the wavelength range below 625 nm for display technology, increasing the visibility for red-blue color-blind people.

References

1. K. V. Chellappan, et al., 'Laser-based displays: a review', Appl. Opt., AO, vol. 49, no. 25, pp. F79–F98, Sep. 2010, doi: 10.1364/AO.49.000F79.
2. T. Nishida, et al., 'Short wavelength limitation in high power AlGaInP laser diodes', High-Power Diode Laser Technology and Applications IX, vol. 7918, p. 791811, 2011, doi: 10.1117/12.872939.
3. D. P. Bour, et al., '610-nm band AlGaInP single quantum well laser diode', IEEE Photonics Technology Letters, vol. 6, no. 2, pp. 128–131, Feb. 1994, doi: 10.1109/68.275404.
4. H. Hamada, et al., 'Room-temperature CW operation of 610 nm band AlGaInP strained multiquantum well laser diodes with multiquantum barrier', Electronics Letters, vol. 28, no. 19, p. 1834, 1992, doi: 10.1049/el:19921170.
5. F. Mauerhoff, et al., 'Optimization of high power AlGaInP laser diodes at 626 nm', Opt Quant Electron, vol. 56, no. 3, p. 419, Mar. 2024, doi: 10.1007/s11082-023-06028-5.
6. F. Mauerhoff et al., 'Continuous wave operation of broad area and ridge waveguide laser diodes at 626 nm', IET Optoelectronics, p. ote2.12125, Jul. 2024, doi: 10.1049/ote2.12125.

Acknowledgements

The work was carried out in the QNC-Space project BaSyQ and was funded by the Federal Ministry of Education and Research within the FMD-QNC project under the funding code 16ME0830. The responsibility for the content of this publication lies with the authors.

MEMS Technology as Enabler: From Accelerometer to LiDAR

Lj. Ristic,[1] V. Milanović,[1] M. Vujosevic,[2] and A. Kasturi,[1]

[1] Mirrorcle Technologies, Richmond, CA 94804, USA
[2] SensSpree, San Jose, CA 95127, USA

ljr@mirrorcletech.com

Abstract. This paper examines the evolution of Micro-Electro-Mechanical-Systems (MEMS) technology by analyzing high-volume MEMS products introduced over the last several decades. It covers MEMS devices widely used in high-volume applications, as well as MEMS mirrors and their role in optoelectronic applications. The discussion of non-optical MEMS applications highlights well-established products such as accelerometers, pressure sensors, microphones, speakers, and silicon timing devices, each shipping in billions of units annually. The focus then shifts to MEMS mirrors, emphasizing their critical role in opto-electronics. Key applications in this space include vector graphics laser projectors (VGLPs), mixed reality/augmented reality (MR/AR) solutions, free-space optical communication (FSOC), MEMS mirror for optical switching for data centers, and MEMS mirror systems for humanoid robots. Finally, special attention is given to the current state of LiDAR development efforts.

Keywords: MEMS Sensors, MEMS Mirrors, LiDAR, VGLP

1 Introduction

The enormous advances in MEMS sensors made over the last couple of decades and the proliferation of numerous sensor products have been primarily driven by progress in silicon micromachining. Ever since Petersen published his seminal paper on silicon as a mechanical material [1] and Howe and Muller developed polysilicon surface micromachining technology [2], leveraging silicon integrated circuit technology for micromechanical structures has been a focal point for sensor researchers, engineers, and startups. For those interested in exploring the use of silicon and polysilicon in the fabrication of micromechanical structures, references [3,4] provide further insights, while reference [5] discusses how the Complementary Metal Oxide Semiconductor (CMOS) process is being leveraged in micromachining.

The worldwide efforts in this field have not only advanced micromachining but have also established MEMS technology as a mainstream technology that has enabled creation of many diverse sensor products, with billions of units shipped annually. In this paper, we highlight some of these sensor products and discuss the future of MEMS Mirrors.

S. Kurimura and T. Yagi (eds.), *Proceedings of Laser Display and Lighting Conference 2025*, Springer Proceedings in Physics 429,
https://doi.org/10.1007/978-981-96-9782-3_17

The paper has two parts. In the first part, we highlight some of the most commercially successful MEMS sensor products, focusing on accelerometers, pressure sensors, microphones, speakers, and silicon timing devices. We describe the key features and technical advances that led to their wide-spread adoption. In the second part of the paper, we explore the optical applications of MEMS Mirrors as a crucial technology in this domain. We discuss several important applications, including VGLP, MR, FCOS, optical switching, and robotics. Finally, particular attention is given to LiDAR and an outline is proposed for its massive adoption.

2 High Volume Applications

The list of MEMS products in high volume manufacturing is rather extensive. However, since the objective of this paper is to highlight exceptionally successful products that have reached mass production, this section will focus only on those that have achieved annual shipment exceeding one billion units. These include accelerometers, pressure sensors, microphones, speakers, and silicon timing devices.

2.1 Accelerometers

Accelerometers were among the first sensors to leverage surface micromachining and successfully be introduced to automotive industry. In the late 1980s and early 1990s, Motorola and Analog Devices led the development of accelerometers for airbag applications. While Analog Devices adopted the comb-structure approach invented by Howe and Lee [6], Motorola pursued a distinct path, developing a three-polysilicon-layer surface micromachining technology and a differential capacitive device as the foundation for its accelerometers [7,8]. Ultimately, both companies were successful in introducing products to automotive customers.

Fig. 1 shows the basic structure of MEMS accelerometer based on differential capacitor developed by Motorola. The device had to satisfy the stringent requirements of the automotive industry: sustaining 1000 g acceleration, incorporating self-test feature and ESD-protection, and being housed in a low-cost hermetic plastic package. All these requirements represented enormous challenges, further compounded by the need to develop a surface micromachining process. Amazingly, all requirements were met. The surface micromachining process with stress-free polysilicon was developed [9], the wafer bonding process was established with a conductive cap [10] to ensure both hermeticity and ESD protection, while a cleverly designed sensing cell took care of self-testing and resilience to high acceleration.

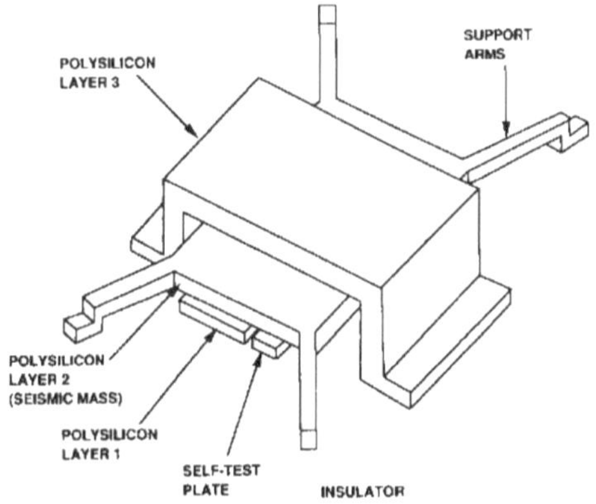

Fig. 1. Polysilicon surface micromachined accelerometer.

The robust capped MEMS accelerometer was ready to be integrated into a plastic package. The final product (Fig. 2) was designed as a two-chip solution where the Integrated Circuit (IC) die accompanied MEMS sensing die in the lead-frame based overmolded package with a built-in stress isolation mechanism [11].

This Motorola's approach enabled a high-yield, high performance, and low-cost solution that was widely accepted by the automotive industry.

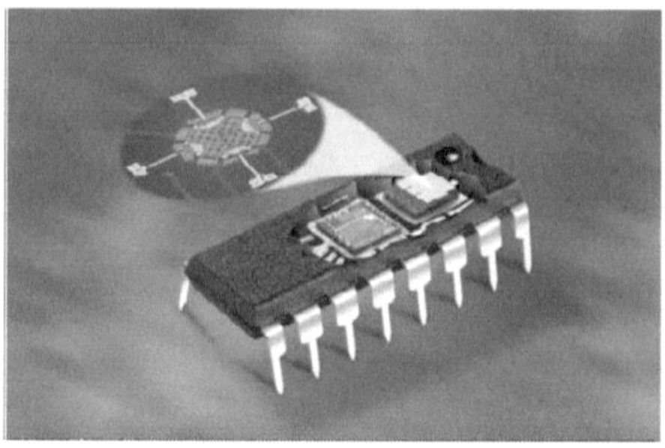

Fig. 2. Motorola accelerometer package for automotive safety application (top mold removed to showcase the IC and MEMS dies, which are mounted side-by-side and electrically connected via wire bonds.) The MEMS die sensing element is also highlighted.

Since the 1990s, MEMS accelerometers have significantly expanded their range of applications. Advances in miniaturization, high-yield manufacturing, and cost-effective production have driven their continued growth in the automotive market and facilitated their widespread adoption across new markets, including consumer electronics, industrial automation, robotics, medical devices, the Internet of Things (IoT), and defense. Today, MEMS accelerometers play a crucial role in applications such screen orientation detection, movement monitoring, step counting, gesture recognition, structural health monitoring, and vibration monitoring of machinery, among many others.

2.2 Pressure Sensors

Pressure sensors are devices that convert mechanical pressure into an electrical signal. A key component of every MEMS pressure sensor is a diaphragm. There are various manufacturing techniques for fabricating silicon diaphragms [3,4,12]. The chosen technique is primarily determined by the sensor's underlying operating principle.

Generally, two fundamental principles are exploited in MEMS pressure sensors: the piezoresistive effect [13] and the capacitive sensing [14]. In the case of the piezoresistive effect, a piezoresistor is embedded within the diaphragm. When the diaphragm deforms due to pressure changes, the resulting strain alters the resistance of the piezoresistor, which is then correlated to the pressure. In capacitive sensing, the diaphragm itself serves as one electrode of a capacitor. Changes in pressure cause diaphragm movement, altering the capacitance, which is then used to determine the pressure variation.

Piezoresistive Pressure Sensors. In this approach, a silicon diaphragm is fabricated in silicon crystal using wet anisotropic etching of a silicon wafer with a (100) orientation. The bulk micromachining process leverages the fact that silicon crystallographic planes with a (111) orientation etch at a significantly slower rate than those with a (100) orientation [15-17]. This results in a characteristic sloped, truncated pyramidal cavity beneath the diaphragm (Fig. 3a).

Piezoresistive resistors are strategically placed at the edges of the diaphragm to maximize sensitivity, as these areas experience the highest stress during membrane bending. The piezoresistor requires a special design, such as the X-ducer used in Motorola pressure sensor [18], or the Wheatstone Bridge configuration used by STM [19]. Once the sensing element is fabricated, the MEMS wafer is bonded to a holder wafer, creating a sealed cavity (reference chamber) for absolute pressure sensing. This approach has been widely adopted by many pressure sensor manufacturers.

The next step is packaging. Unique packaging schemes for pressure sensor devices enable a wide range of applications, from absolute pressure measurement to differential pressure sensing. Over the years, MEMS pressure sensor packaging technology has evolved, developing various types of packaging schemes to protect the die while allowing exposure to the environment. In addition to protection from thermomechanical stresses, specific shielding features against chemically aggressive environments have been critical for expanding into new markets, including automotive, industrial, consumer, and medical [20]. The complexity increases for certain applications that require specialized materials — for example, white plastic based polysulfon is used in medical devices for its chemical resistance and biocompatibility [21].

Capacitive Pressure Sensors. In the case of capacitive pressure sensor, a surface micromachining is used in combination with a thin polysilicon layer and sacrificial etching. The sensor features a sealed cavity and a thin diaphragm, which also functions as one of the capacitor plates (Fig. 3b). The pressure sensor sits on the top of the MEMS die. This technology is extensively used for making high-pressure sensors. Capacitive pressure sensors form the core of Tire Pressure Monitoring Systems (TPMS) [22]. A notable example is Freescale's TPMS solution, which integrates a capacitive pressure sensor, accelerometer, temperature sensor, integrated MCU, 16 kilobytes of RAM, low-frequency receiver, and high-frequency transmitter, all in a Single inline Package (SiP) [23]. Accelerometer turns off the device when vehicle is still, but also records acceleration when moving. Pressure sensor measures tire pressure, while temperature sensor records temperature for data correlation. LF receiver serves for programing and resetting, while HF transmitter transmits data for warning and data analytics. This product has become a cornerstone in the automotive industry, widely used in both passenger vehicles and commercial trucks for real-time tire pressure monitoring.

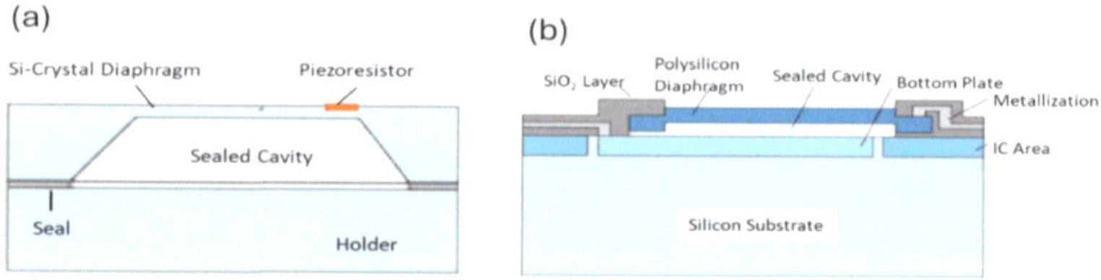

Fig. 3. (a) Piezoresistive pressure sensor fabricated using wet anisotropic etch; (b) Capacitive pressure sensor fabricated using surface micromachining.

2.3 Microphones

A MEMS microphone is an acoustic sensor that converts sound into an electrical signal. There are different types of MEMS microphone technologies [24], but the most dominant type is capacitive MEM microphone which account for 80% of the market. The basic structure of a capacitive MEMS microphone is shown in Fig. 4. [25]. It consists of a very thin diaphragm and a backplate. The sound wave creates air pressure that moves the diaphragm relative to the fixed backplate. The resulting change in capacitance between the diaphragm and the backplate is converted into an electrical signal and processed by an ASIC. The ASIC die is a separate die that, together with the MEMS die, is placed within the same package.

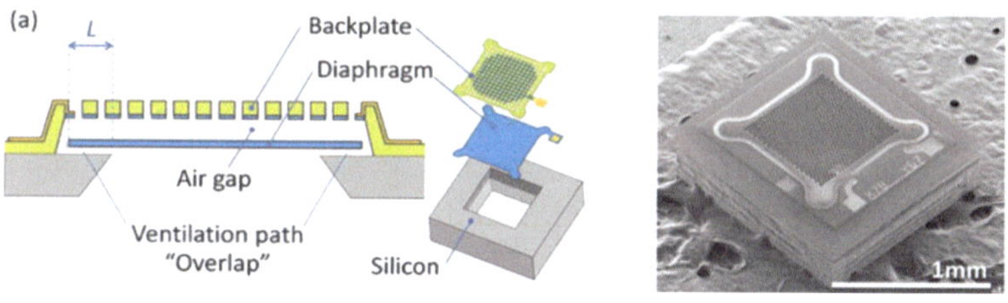

Fig. 4. MEMS microphone: (a) cross-section, (b) SEM image.

A MEMS microphone module is an open-cavity package in which MEMS die and ASIC die are placed side by side on an organic or ceramic substrate and connected via wire bonding. The package includes a sound port that allows sound to enter. This port is often located in the package substrate creating a bottom-port configuration (Fig. 5) [25], but it can also be placed in the lid for a top-port design. The port's location is determined by the integration and placement requirement of the MEMS microphone module within the final system.

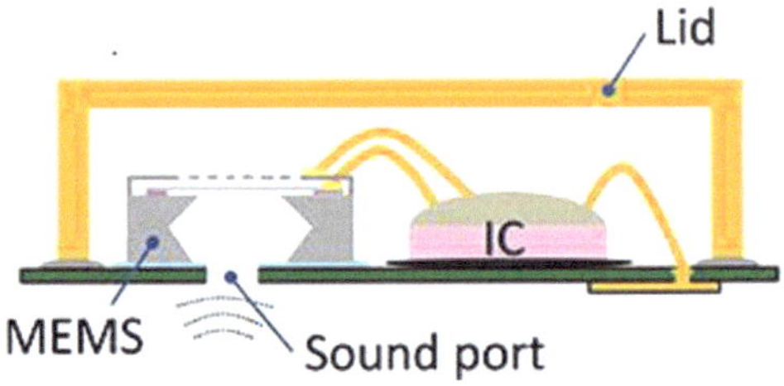

Fig. 5. Microphone module – schematic representation.

Microphones have significantly benefited from ongoing advancements in MEMS technology, leading to substantial miniaturization of these devices. As a result of their small size and low power consumption, MEMS microphones are well-suited for applications in compact, battery-powered devices such as smartphones, tablets, smartwatches, earbuds, hearing aids, laptops, smart speakers, etc. Additionally, new use cases for these devices have very recently driven the development of more environmentally robust technologies, as discussed in [26-28]. The maturity of the technology and its low cost have made MEMS microphones ubiquitous. Moreover, this advancement has transformed how people interact with modern devices, as evident by the widespread adoption of voice control and hands-free communication.

2.4 Speakers

A MEMS speaker produces sound by converting an electrical audio signal into a sound wave using MEMS technology. At its core, the speaker features a diaphragm whose vibrations move the air in the air chamber creating sound. High quality sound generation requires significant movement of the diaphragm. Unlike MEMS microphones that necessitate a very small diaphragm displacements, MEMS speakers require larger diaphragm motion to effectively move air and produce sound.

There are four basic transduction principles used to drive the diaphragm: piezoelectric [29], electrodynamic [30], electrostatic [31], and thermoacoustic [32]. An excellent review of these driving mechanisms can be found in [33]. Currently, piezoelectric MEMS speakers dominate the market. Therefore, as an example, we here briefly describe only the piezoelectric MEMS speaker (Fig. 6) [33].

MEMS speakers are typically designed to operate within a frequency range of 20 Hz to 20 kHz, aligning with the human hearing range. However, in daily life, most audible sounds range from 100 Hz to 10 kHz, including speech (primarily 300 Hz–3.4

kHz) and musical harmonics (extending beyond 6 kHz). Due to the small size and limited diaphragm movement of MEMS speakers, their low-frequency response is often a challenge. Therefore, it is a must to evaluate MEMS speakers across both low-frequency and high-frequency bands.

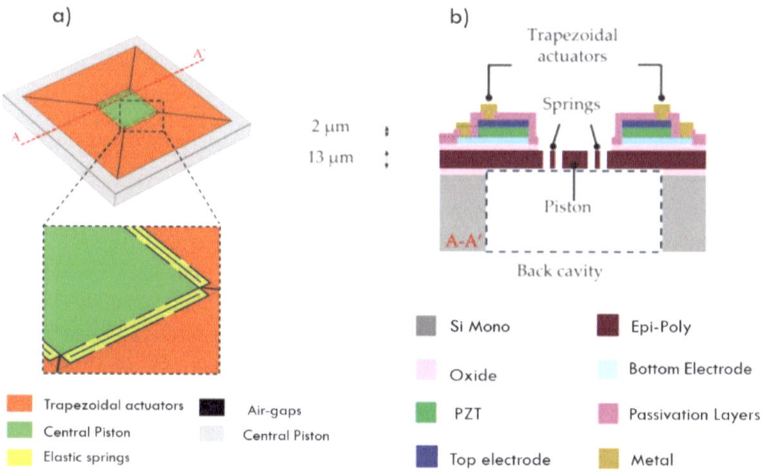

Fig. 6. a) Top view of speaker diaphragm, b) Cross-section of speaker diaphragm.

MEMS speakers are among the most recent product families to leverage MEMS technology, with commercial products emerging only over the last decade. Their development has generally been more challenging compared to MEMS microphones. This challenge is primarily driven by the need for large diaphragm displacement, as mentioned earlier. On one hand, sufficient power is required to generate enough force to move the diaphragm, while on the other hand, the structural integrity of the diaphragm must be preserved despite large displacements and sustained high stresses.

The advantages of MEMS speakers over traditional non-MEMS technologies (such as electrodynamic and balanced armature speakers) are their small form factor, lower power consumption, batch fabrication, and ease of integration with electronics. These benefits make them highly attractive for many applications in consumer electronics and other size- and power-constraint applications.

2.5 Silicon Timing Devices

The main purpose of MEMS timing devices is to provide clock functions that are required in modern electronic products. The frequency range of these devices spans from Hz to MHz and they offer programmability. They are an excellent alternative to crystal oscillators and are gaining acceptance in many market segments including automotive, aerospace and defense, telecommunications, IT, consumer, and medical applications. They offer reliable performance, low power consumption, high stability, small form factors, and low cost.

There are three main categories of MEMS timing devices: resonators, oscillators, and clocks [34,35]. Typically, silicon resonators are designed as devices that serve as the source of frequency, while oscillators include resonator and ASIC to drive it and compensate performance of resonator and make it continuous and stable. Finally, clocks often combine multiple oscillators to create source of multiple reference frequencies.

All commercial silicon MEMS resonators are based on electrostatic driving mechanism. A cross-section example of a resonator developed by Si Time is shown in Fig. 7. [36]. The resonator is etched in silicon on the insulator layer. Since silicon is a stable material, the resonator exhibits high quality factors and near zero frequency drift. The resonator is sealed at the wafer level [36,37] using high-temperature polysilicon deposition to create a clean vacuum environment needed for stability.

The resonator is then assembled in a single package together with a COMS ASIC to form an oscillator. The resonator and ASIC are integrated as a die stack within a plastic package. The ASIC itself represents an art on its own and includes resonator sustaining circuit, charge pump, fractional-N PLL, temperature sensor with digital temperature compensation, non-volatile memory, as well as output dividers and drivers. The final result is an outstanding performance 20 MHz MEMS oscillator with a stability of ± 25 ppm [36].

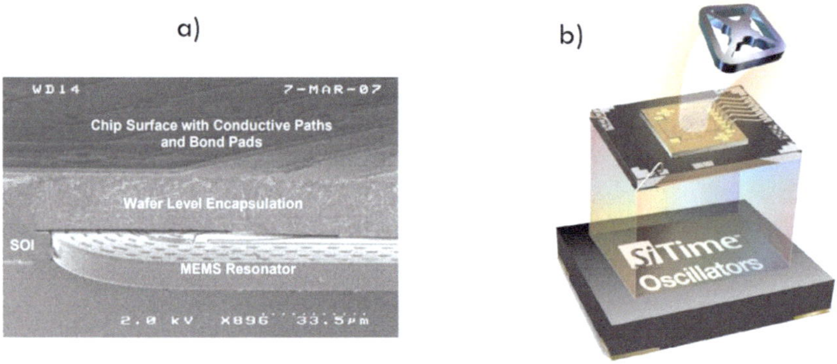

Fig. 7. a) Silicon MEMS Resonator, b) Silicon MEMS oscillator.

3 MEMS Mirror – Essential Device for Photonics Applications

MEMS Mirrors are small silicon-based devices capable of tilting along a single axis (one-dimensional mirror) or in two axes (two-dimensional mirror) [38,39]. Depending on the design they can also move along a third axis making a piston action [38]. Usually, MEMS mirrors are made in a circular shape with diameters ranging from 0.5 mm to 10 mm [40].

The technology for making MEMS mirrors has significantly matured over the last two decades. Today, three primary actuation principles are used to make MEMS mirrors move: electrostatic, electromagnetic, and piezoelectric actuation. A comparative summary of these approaches is shown in (Table 1). Each method has its own advantages and limitations, determining their suitability for different applications.

Silicon is the core material used for making MEMS mirrors across all three actuation techniques, though some methods require additional material to enable mirrors movement. For example, electromagnetic actuation requires use of additional magnets, while piezoelectric driving requires lead zirconate titanate (PZT) films. Designers must make trade-offs between key performance parameters to select the most suitable MEMS mirror for a given application.

Table 1. A comparative summary of basic parameters between electrostatically driven, electromagnetically driven, and piezoelectrically driven MEMS mirrors.

Parameters	Electrostatic Driving	Electromagnetic Driving	Piezoelectric Driving
Core Material	Si	Si	Si
Additional Material	n/a	external magnets	PZT + passivation
Power Consumption	< 1 mW	~ 500 mW	~100 mW
Drive Voltage	high	low	high
Drive Current	very low	high	medium
Temperature Sensitivity	low	high	high
Hysteresis	none	medium	high
Deflection Angle (linear)	(+/-) 7.5 deg	(+/-) 15 deg	(+/-) 15 deg
Device Manufacture	standard IC process	IC + added complexity	IC + added complexity
Electronic Control	simple	medium	closed loop required
Non-resonant Mode	yes	yes	no
Resonant Mode	yes	yes	yes
Vector Graphic Mode	yes	yes	yes
Size	smallest	largest	medium

Examples of different MEMS mirrors are shown in Table 2 [41]. The device diversity offered in various sizes and packages demonstrate that MEMS Mirror technology has reached a high level of maturity.

Table 2. Dual-axis Quasistatic MEMS mirrors ranging from 2 mm to 6.4 mm in diameter.

		 Compact package, <0.4g mass		 Hermetic package, 2.5g mass
MEMS Mirror	A7M20.2-2000AL	A7B1.3-2400AL	A8L2.2-4200AL	A5L3.3-6400AL
Package and Cover	TINY48.4-NW	LCC20-TW	TINY48.4-NW	TO8-CAPSC1
Mirror Diameter	2.0mm	2.4mm	4.2mm	6.4mm
Optical Angle	±10°	±16°	±11°	±5°
First Resonance	1300Hz	600Hz	450Hz	300Hz
Feature	high bandwidth	large angle	large angle*diameter	large diameter

3.1 MEMS Mirror Basic Function and Scanning Engine

A MEMS mirror's primary function is to deflect a focused beam of light in different directions. The light beam can be steered along a single axis (one-dimensional mirror) or along two axes (two-dimensional mirror). This ability to receive and redirect a focused beam of light is fundamental to scanning technology.

In an electrostatically driven mirror, tilting occurs when a voltage is applied to a set of comb-finger electrodes. A 1D mirror has one set of electrodes on opposite sides of the mirror, while a 2D mirror has two sets of electrodes for independent movement. along both axes. Fig. 8 shows MEMS Mirror with two pairs of comb-drives [38], one pair for each axis. These electro-static actuators (comb-drives) are an integral part of the MEMS mirror device, whereas the electronic circuit that controls the tilting is a separate IC.

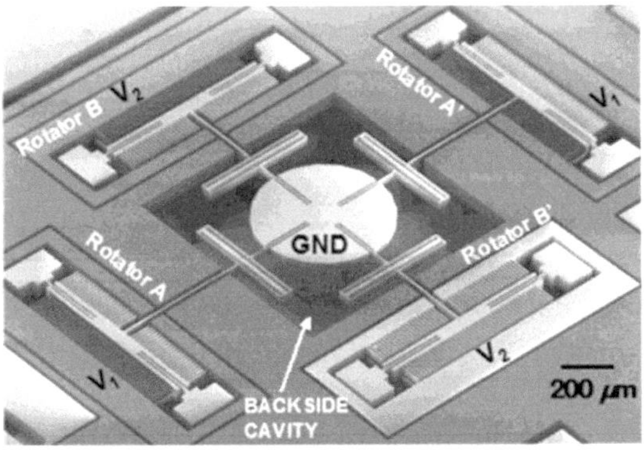

Fig. 8. SEM of an early demonstration of 2D-MEMS Mirror with two pairs of comb-drives and gimbal-less structure allowing efficient dual-axis tip/tilt as well as piston actuation. Comb-drive pair A-A' tilts mirror along x-axis, and pair B-B' tilts mirror along y-axis.

The device architecture shown above is the essential modular building block for a portfolio of MEMS mirrors with optimization for different use cases and applications. The architecture covers cases for single- and dual-axis actuation, resonant and broadband (quasi-static) actuation, as well as different mirror sizes and coatings.

The scanning capability of MEMS mirrors is cleverly utilized to enable two basic functions for manipulating laser light: directed light projection and directed light acquisition (imaging). There are applications that utilize either functionality and some applications in fact utilize both projection and imaging (e.g. LiDAR). In the following sections we describe multiple applications of MEMS mirror-based systems, and then we pay special attention to LiDAR technology.

3.2 Vector Graphics Laser Projector and Applications

Laser projectors convert moving light into higher-level information such as signs, graphics, and videos. Visual information can be presented using several methods, including Vector Graphics Laser Projection (VGLP) mode, raster mode, and Lissajous mode. MEMS-based VGLP (MEMS VGLP) is currently a topic of great interest due to its ability to produce high-brightness images even in outdoor daylight conditions. In this paper, we focus exclusively on the MEMS VGLP method.

MEMS VGLP enables compact pocket-size projectors capable of providing high brightness visual messaging in daylight environment on multitude of different surfaces. The key differentiation compared to other projection technologies is its VGLP concept, which enables superior high-brightness performance. A distinguished product in this category is branded as Playzer by Mirrorcle Technologies [42,43]. A high-level block diagram of this device is shown in Fig. 9.

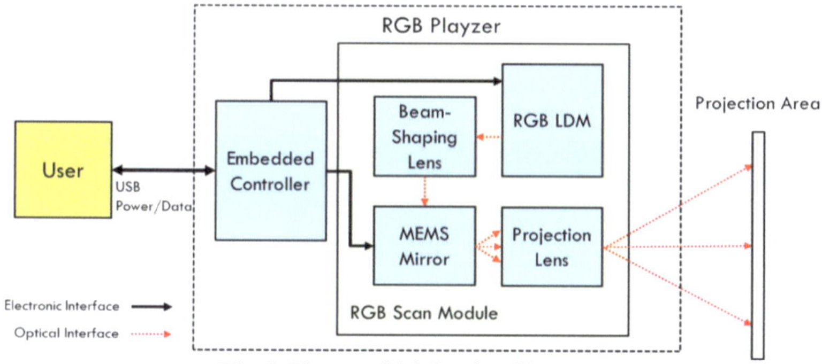

Fig. 9. Playzer - high level architecture.

The hardware consists of a MEMS mirror, one or more lasers, optics, and electronics. An essential part of this product is its software. The RGB laser diode module (RGB LDM) serves as the light source, generating one or more laser beams that are optically focused by a beam-shaping lens and directed toward the MEMS mirror. The mirror steers the beam in different directions to create the desired message in the form of graphics or text. Both the RGB LDM and the MEMS mirror are controlled by integrated electronics and software.

Monochrome Playzer version has been implemented with different laser wavelengths, particularly in four different colors (red, green, blue, and violet). Laser intensity or brightness of the vector graphics is controlled by a laser driver which allows 256-levels of control within content data but also has programmable gain and offset to calibrate to each laser diode and modulate based on environmental conditions. On the other hand, RGB Playzer includes 3 laser sources (red, green, blue) and is a full color projector (256 x 256 x 256 colors). It is integrated into a small box, approximately 90 cm3, weighing under 200g, and consuming less than 2W of power. Notably, the majority of the power is consumed by lasers. The Playzer is programmable, allowing for the creation of customized messaging tailored to specific applications. Several examples of the Playzer in action are shown in Fig. 10. Fig. 10a is an example of MEMS mirrors used for video projection at video-raster rates (720p in this example). This produces color-rich imagery and videos, however with very limited brightness which practically limits the use of technology to dark environments and short distance projection. In contrast Fig. 10b shows RGB Playzer with same RGB laser capability and similar MEMS Mirrors but due to the VGLP nature of projection it is projecting bright content at a building wall more than 10m away and during daytime (shaded side of building). Other examples illustrate various automotive applications of MEMS mirror based VGLP such as head up display (HUD) in Fig.10c, programmable courtesy lighting in Fig.10d and 10e, and programmable headlights (prototype dynamic laser lighting projector in Fig. 10f projects programmable white vector content onto a building at >10m distance).

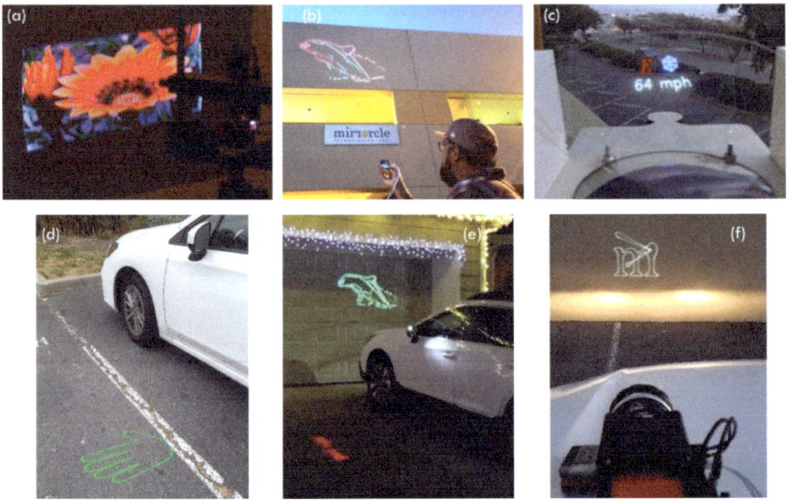

Fig. 10. a) MEMS raster projector, b) RGB Playzer in VGLP mode, c) VGLP HUD display, d) VGLP courtesy lighting – projection om the concrete, e) VGLP courtesy lighting projection on the wall, f) dynamic laser light projector – white VGLP projection on a wall

Applications for VGLP are numerous and include automotive, agriculture, robotics, MR, commercial, consumer, industrial, transportation, and smart city markets.

3.3 Mixed Reality (MR) Applications

Augmented Reality (AR) is a term that began gaining traction in the late 1990s [44,45] to describe systems designed to enhance user perception of the real world by supplementing it with 3D virtual objects that coexist within the same space as the physical world. The existence of virtual objects led to the term Virtual Reality (VR), while the blending of real and virtual objects gave rise to the term Mixed Reality (MR). These terms are often used interchangeably to describe systems that merge with real and virtual environment and allow for real-time interaction. In this paper, we will use the term *Mixed Reality (MR)*.

It is important to note that MR is not limited to specific display technologies or to visual perception alone —it applies across all human senses. Over the past two decades, there has been enormous effort and investment in developing MR technology and products. These efforts can be broadly categorized into two major areas: 1) Immersive MR, featuring head-mount displays (HMD), which are designed to provide full immersion with a rich fusion of the real and virtual worlds, and 2) Informative MR, with smart glasses as a representative product, prioritizing simplicity usability and cost.

While considerable progress has been made, HDMs have yet to achieve the goal of full immersion as originally envisioned. Smart glasses, on the other hand, are more practical in their approach, focusing on simplicity, usability, and affordability.

MEMS Mirrors play a crucial role in imaging formation for both categories of MR products. They are used to deliver images (foveated rendering) to each eye separately, which means that at least two mirrors are required per product. Additionally, MEMS mirrors are used for eye tracking and 3D scanning, meaning that some HMD products may incorporate up to eight MEMS mirrors. Fig. 11 provides an example of smart glasses designed by Bosch [46]. Fig. 11a illustrates how the temple of the smart glasses is utilized to integrate the entire electronics. Fig. 11b shows the temple hinge where the MEMS mirror is placed, while Fig. 11c depicts the lens of the glasses, where the visual information is displayed. In this design, four one-dimensional mirrors are used per pair of smart glasses.

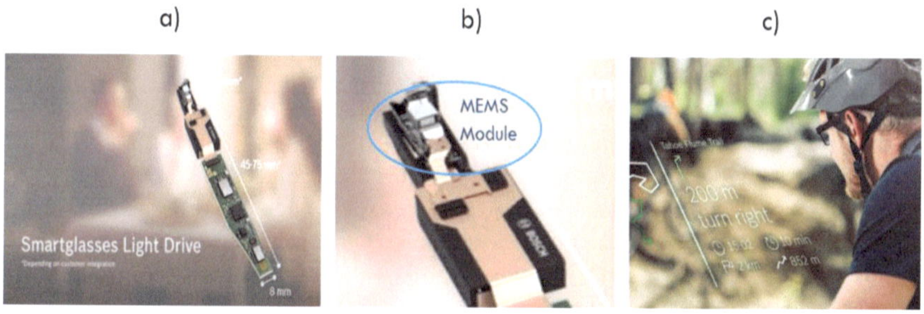

Fig. 11. a) electronics is incorporated in temples, b) MEMS mirror module is located at the hinge of the temple, c) smart glasses use lenses for smart messaging display.

3.4 FSOC Applications

Free space optical communications (FSOC) use high-frequency modulated optical beams for sending data back and forth between optical transceivers [47]. Line of sight (LOS) is essential for FSOC. This technology offers high transmission rates (large bandwidth), lower power consumption (almost half that of RF communication technology), and reduced equipment dimensions (1/10th of the diameter of an equivalent distance/bandwidth RF antenna or even smaller) [48,49]. FSOC is easy to install and license-free, unlike RF standards. This also relates to the fact that the communications are secure as they are directed and broadcasted only toward intended recipient and difficult to intercept. Because of these advantages, it is one of the front runners for broadband link deployment in many space and terrestrial use cases. Typically, FSOC operates in the infrared range at ~1520nm-1560nm due to better eye-safety, and possibility to operate laser at higher power. It should be noted that early development of FSOC began more than 50 years ago, for applications in the space and defense industry in the US, and in the recent 10 years MEMS mirrors have often become integral part of the system solution [50] as they enable significant overall reduction in system size and cost.

While FSOC is typically focused on space (satellite-to-satellite, satellite-to-earth) applications, and more recently drone applications, there is also strong interest in terrestrial applications. Namely, FSOC is considered as a good solution for the last mile communication in urban areas [47,51]. Additionally, FSOC can be used for connections of Local Area Networks (LANs), Metropolitan Area Networks (MANs), or to support backhaul network for mobile wireless and optical fibers networks. Unfortunately, FSOC also has disadvantages, the most significant being its susceptibility to bad weather conditions, which can degrade the communication link [51]. The FSOC laser signal may get absorbed and scattered by tiny particles, such as aerosols, smoke, dust, and haze [52]. However, practical data indicate that actual down-time is only a few percent of the time, which is acceptable for many difficult to access areas lacking communication infrastructure.

A few examples of successful developments and applications as well as rendering of some applications are shown in Fig. 12. An excellent example is MEMS mirror based compact laser terminal developed by Deutsches Zentrum für Luft- und Raumfahrt, shown in Fig.12b, that was used in pioneering launch of a compact satellite [53]. Another good example would be the FSOC success is GoogleX's Taara project [54, 55]. GoogleX has developed Project Taara as a high-speed cost-effective FSOC system for installation in underdeveloped areas. It can be deployed in rural and difficult to access areas, as well as in urban areas as a last mile extension. The range of the system is 20 km, and it provides speeds of 20 Gbps. Other advantages are low cost, fast and easy deployment, and low environmental impact. Taara products have been already installed in 13 countries. Fig. 12.c shows installation of a Taara FSOC transceiver.

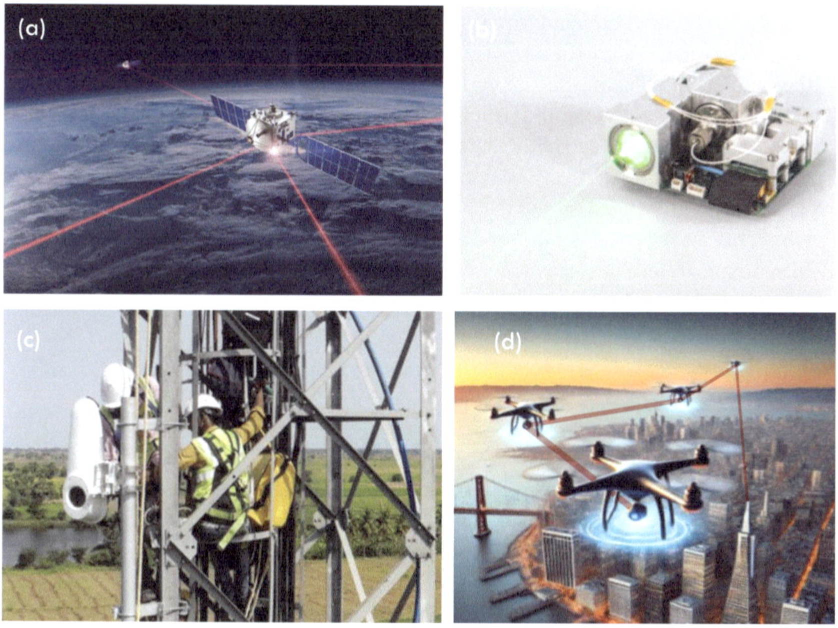

Fig. 12. Examples of FSOC applications with laser beam steering solutions: a) rendering of satellite-to-sattelite FSOC use case with multiple transceivers per satellite, b) MEMS mirror based compact laser terminal by Deutsches Zentrum für Luft- und Raumfahrt , c) Taara FSOC terminal installation, and d) rendering of drone-to-drone and drone-to-infrastructure FSOC application.

3.5 Optical Switching Applications in Data Centers

Artificial Intelligence (AI) will be one of the key driving forces in the next decade, fueling unprecedented demand for data centers and the infrastructure networks that connect them. One of the primary requirements for efficient AI is a development of sophisticated learning models fueled by large datasets. These models demand seamless, simultaneous connectivity across numerous data centers without dropping a single link. This, in turn, necessitates the deployment of ultra-reliable, high-speed switching networks.

Optical switching is emerging as one of the key technologies to revolutionize data networks. As opposed to traditional electronic switching, optical switching eliminates the need to convert light into electrical signal and then back into light, thereby reducing latency and energy consumption. This makes it particularly promising for AI-driven applications and hyperscale datacenters, where massive amount of data requires seamless, instantaneous connectivity, and the energy demand of data centers is a significant concern.

The concept of optical switching using MEMS Mirrors has been described by multiple authors [56,57] more than 20 years ago. Most recently Google has successfully commercially adopted optical switching with what they call Optical Circuit Switching (OCS) [58]. A MEMS mirror array is at the core of this technology, allowing hundreds of fiber cables to be connected directly to each other without the need for optical-to electrical-to optical conversion. Here MEMS mirrors simply redirect the optical signal, eliminating the need to decode the data packet. Fig.13 illustrates the basic principle of switching between arrays of MEMS mirrors used in the data center network. However, several challenges related to yield, space and cost remain when using Google's approach. We have proposed a modified method utilizing 3D heterogenous integration of MEMS mirrors to form an array [59], which would further improve yield, reduce space, and lower power consumption.

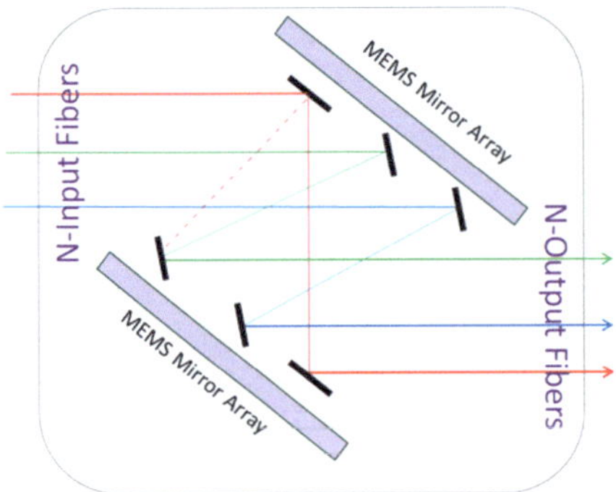

Fig. 13. Optical cross-connect switching using array of MEMS mirrors.

Additionally, a wireless version of optical switching based on MEMS mirrors has been demonstrated for interconnecting servers in data centers [60]. Based on these demonstrations and already commercialized application mentioned earlier, MEMS mirrors are expected to play a center role in optical switching networks for data centers.

3.6 Robotics Applications

Robots and drones are changing our lives and transforming the entire industrial and service sectors, including manufacturing, transportation, automotive, health, and agriculture segments. The integration of robotics in many of these areas improves operations and reduces operating costs. A massive deployment of autonomous mobile robots and drones improves space optimization, efficiency, process control, as well as delivery. At the same time, the introduction of robotics in the work environment creates a new dynamic between the human workforce and machines. Interaction and intention communications between them are frequently required and absolutely critical for safety.

Existing human-machine interaction solutions in the working environment are often based on sound or simple flashing light signals [61]. However, working conditions in manufacturing facilities and warehouses are typically very loud and noisy, and on top of that, they often employ bright light. This significantly hinders human-machine interaction based on sound or inefficient light messaging. An alternative approach based on MEMS VGLP is proposed in [62]. As shown in Fig. 14, programmable visual messaging can be used for efficient interaction, whereby custom messaging can be clearly displayed. This enables the robot to signal its intentions for its next move or communicate instructions to humans. The proposed VGLP solution is optimized for robotics, offering low power consumption, low weight, small form factor, and high-contrast messaging, outperforming any other existing technology.

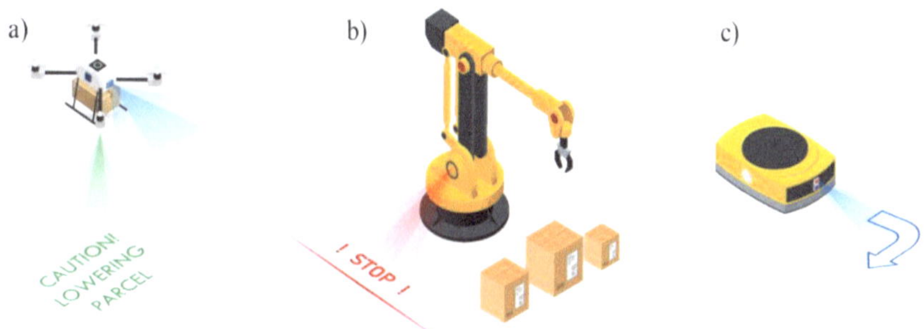

Fig. 14. Programable visual messaging in robotics. a) drone signals intention, b) robot warns for not coming closer, c) robot vacuum cleaner signals intention to move left.

3.7 LiDAR Technology: A Possible Path to Mass Adoption

LiDAR (Light Detection and Ranging) systems have been in the spotlight over the last decade primarily due to the hype surrounding autonomous driving and the expectation that they would become one of the crucial enablers of that technology. However, the high expectations, delays in the widespread adoption of autonomous driving, and inadequate maturity of LiDAR technology have led to disappointment, prompting a reevaluation of LiDAR in the industry.

In this section we provide a high-level analysis of LiDAR technology and outline a potential path to future success.

Why LiDAR in the first place? LiDAR is an active device that senses its surroundings by emitting laser light into an area of interest and measuring the distance to objects in the area of interest. It does so by analyzing the laser beam reflected from the object [63]. LiDAR can generate a three-dimensional representation of its surroundings, hence the term 3D perception sensing, by steering the laser beam in two dimensions and capturing distance or range data (third dimension). LiDAR is more precise than radar in object detection and has an advantage over cameras, as cameras typically lack depth-sensing capabilities. However, LiDAR also has drawbacks – its performance deteriorates in bad weather conditions (rain, snow, fog, dust).

LiDAR Architecture. The high-level architecture of LiDAR is shown in Fig.15. The first insight is that LiDAR is a rather complex system consisting of multiple distinct blocks. The selection of each component plays a critical role in determining both the system's final performance and its overall cost.

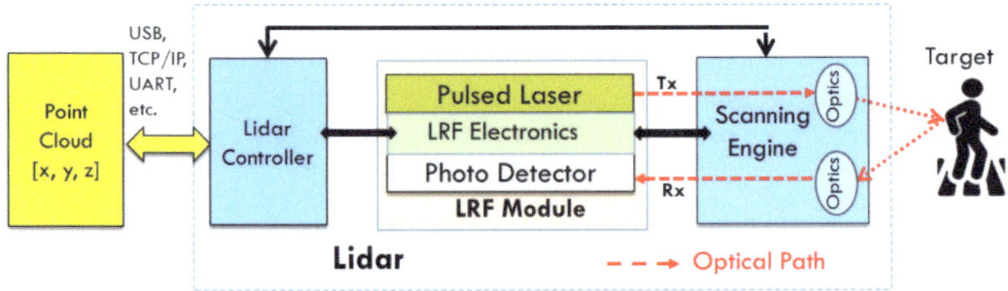

Fig. 15. High-level LiDAR Architecture.

The critical blocks of a LiDAR system are laser source, scanning engine, photo detector, LRF electronics (laser range finder), and optics. The basic principle of operation is as follows: The pulsed laser sends a transmitting beam via the scanning engine and transmission optics into the surrounding space. When the laser beam hits an object, it reflects back and, via the receiving optics, it reaches the photo detector. The number of photons detected depends on the distance. The electronics compare the transmitted and received signals and didact information about the distance. That information, combined with the scanning position, gives a point cloud. The entire process is repeated at scanning rates from Hz to MHz to create a 3D perception image, that represents the scanned environment. Let us now examine the critical functional blocks, starting with lasers.

Laser Source and Wavelength. A laser beam is emitted by a device known as a laser diode, often simply called laser. These lasers are typically classified into two groups based on their lasing mechanism: edge emitter laser (EEL) and vertical cavity surface emitting laser (VCSEL) [64]. EELs emit a laser beam in an elliptical shape from the edge of device, whereas VCSELs emit a circular-shaped laser beam from the surface of the device. EELs are a more mature technology and have been in use for decades. VCSELs, on the other hand, are a newer technology but offer easy creation of two-dimensional array on a single chip, and that is how they are generally used. Both types of lasers operate in pulsed mode. VCSELs typically consume higher power in LiDAR applications compared to EELs.

Lasers used in LiDAR operate in infrared wavelength regions. The most commonly used wavelengths today are between 850 nm and 905 nm, classified as Near Infrared (NIR), and 1550nm, which falls into the Short-Wave Infrared (SWIR) category. NIR lasers are more technologically mature than SWIR lasers. On the other hand, NIR lasers are closer to the visible light spectrum and more dangerous to the human eye when operated at high power. In contrast, 1550 nm lasers, being father from the visible spectrum, can be operated at higher power levels with lower eye safety concerns. When selecting a wavelength, one must also consider laser performance under adverse weather conditions [65,66]. The presence of particles in the air scatters and absorbs photons and degrades LiDAR performance. NIR lasers are less affected by adverse weather conditions compared to lasers operating at 1550nm wavelength.

Photodetector. A photodetector is a device that converts optical signals into electrical signal, through a phenomenon known as the photoelectric effect. Photosensitivity is very important characteristics of photodetectors, and it depends on the wavelength; therefore, the choice of photodetector is closely related to the choice of laser. An excellent review on the subject is presented in [67]. The most common photodetectors used in LiDAR are Avalanche Photo Diode (APD), Single-photon Avalanche Diode (SPAD), and Silicon Photomultiplier (SiPM). APDs have a current-gain of around 100, low noise, and they are well established, making them widely used in LiDAR systems. The photosensitivity of silicon APDs spans through the visible region of up to 1000 nm, making them unsuitable for 1550 nm wavelengths. To pair them with SWIR lasers one needs to choose indium gallium arsenide (InGaAs) APDs, which, while effective, are more expensive than silicon APDs.

SPADs are popular due to their ability to detect only a few photons within a short period of time. They have an exceptional gain in the order of 10^6, allowing them to detect weak photo signals over long distances. Since they are manufactured using the CMOS process, it is easy to design SPAD array on a single chip.

SiPMs operate on the same principle as SPADs do but can count photons and provide information on the magnitude of the photon flux. This is achieved by designing a dense array of microcells, each composed of paired SPADs. Because of the increased complexity, they are more expensive.

LRF Electronics. Laser Range Finder is a method used to measure distance to an object by using a laser beam. The way the laser beam is modulated defines the type of electronics. Two methods are most commonly used today: 1) Time of Flight (ToF) Method [68], also known as direct detection and 2) Frequency Modulated Continuous Wave (FMCW) Method [69], also known as coherent detection.

ToF Method uses a pulsed laser to determine the distance to an object. It predominantly uses NIR wavelength and is a simpler method between the two in terms of structure and signal processing. Because of the simplicity, ToF is the dominant technology in the market today. However, to increase the range in ToF system requires increasing the laser power, which is not desirable due to eye safety concerns. So, there is a tradeoff that needs to be made.

FMCW operates by constantly emitting frequency-modulated laser signal toward an object while maintaining a reference signal locally (as a local oscillator). The reflected signal from the object is then mixed with the local oscillator signal, producing an intermediate frequency. Using the intermediate frequency and the shift in frequency due to Doppler effect, both the distance and velocity of the object can be simultaneously determined. FMCW is a more complex method, but it offers the advantage of providing additional information on the velocity of the targeted object.

Scanning Engine. The Scanning Engine, also called the Scanning Method, describes how the transmitting laser beam is steered to explore the area of interest. Scanning engines can be classified into mechanical and solid-state types.

Mechanical scanning techniques [70] use bulky rotating mirrors to steer the laser beam through rotation. The most popular mechanical systems use spinning mechanisms with polygonal mirrors and motors. While these systems are bulky and prone to vibration-induced failures, they provide a full 360^0 horizontal field of view.

Solid-state scanning refers to devices used for laser steering that are embedded in the silicon chip. The most common solid-state method involves using a MEMS mirror for laser beam steering. In this case, the MEMS mirror steers the beam following a programmed scan trajectory. Advantages of MEMS-based LiDAR include a small form factor, low power consumption, high reliability of silicon MEMS mirrors, and the ability to dynamically change the field of view, as we have discussed in [71]. This reference also provides an extensive overview of various LiDAR architectures and discussed various trade-offs and uses of MEMS Mirrors.

Flash scanning [72], another solid-state technique, uses a combination of a single laser and an optical diffuser to illuminate the entire scene of interest at once. This technique requires a corresponding photodetector, consisting of a two-dimensional array of photodiodes to capture reflecting returns. In this case, a major drawback is the high laser power required to illuminate the entire scene simultaneously. As a result, flash LiDAR is mainly suitable for short range detection of less than 10m. Another limitation is the restricted field of view defined by optical diffuser. On the positive side, flash scanning eliminates challenges associated with moving laser beam.

Additionally, there are emerging scanning techniques such as optical phase array (OPA) [73,74] where steering of the laser beam is achieved through phase modulation. However, this technology is still far away from maturity, and it will require some time before gaining traction.

Optics. Optics systems are made of lenses, and they play an important role in LiDAR architecture - many would say a crucial role, and they are correct. Lenses are used on both sides of LiDAR: on the transmitter side and on the receiver side. It is important to understand that lenses need to be tailored to specific requirements that depend on the architecture of the LiDAR [75]. Advanced technology plays a prominent role in designing and manufacturing optics for LiDAR. CAD tools such as ZEMAX are frequently used [76], along with advanced materials and coatings, to achieve the desired field-of-view (FOV) and performance with minimal distortion.

The Specific requirements imposed on lenses stem from the function they play in the LiDAR architecture. For example, in ToF LiDAR, the role of the lenses on the transmitter side is to focus the laser beam and keep it collimated (maintaining the beam diameter over distance with minimal divergence), while the role of the lenses on the receiver side is to collect as many photons as possible and focus them onto photodetector. In the case of Flash LiDAR, the lenses on the transmitter side are used to distribute light uniformly over the entire field of view, ensuring uniform illumination of the scene, while the lenses on the receiving side are optimized to focus reflected light onto the array of receiving detectors, enabling a high resolution image from the scene.

In summary, lenses for LiDAR play a critical role in defining FOV. They must be customized to the requirements of the particular architecture. To achieve the desired performance, they may incorporate aspherical surfaces, diffractive elements, or multi-element configurations optimized for specific scanning patterns. All of this underscores that optics indeed plays the crucial role in the LiDAR design.

Software and Data Processing. In LiDAR systems, software and data processing are equally important as hardware. The software synchronizes laser operation with the scanning engine while also managing the scanning pattern. Once data is obtained from the reflecting laser beam, the software processes the data to generate a 3D point cloud or a 2D image representing the scanned area. It often uses advanced algorithms to enhance the data output by filtering noise and correcting distortions, atmospheric effects, and drift over time and temperature. It also offers programmability in operation due to its ability to zoom-in to specific areas of interest. A good example of custom software developed for Lidar is shown in [71].

LiDAR is often integrated with sensing systems such as camera, radar, inertial sensors, or magnetic field sensors, in which case the software fuses data to augment the quality of results (see the section on data fusion). Finally, with the rise of AI, it is expected that machine learning and other smart algorithms will play ever increasing role in interpreting the LiDAR data.

LiDAR's path to mass adoption. Having reviewed the various functional blocks and technologies involved in building a LiDAR system, an important question remains: What is the optimal path to enable the massive adoption of LiDAR? As always, the answer lies in achieving an acceptable cost while maintaining satisfactory performance.

Can we bring the cost down to between \$50 to \$100? Can power consumption be brought down to approximately 1W? Can the system be miniaturized to matchbox size and weigh less than 50g? We believe the answer is emphatic YES to all these questions. While such a solution may not meet the requirements of every application, it would certainly be suitable for many.

Let us outline a viable path toward low-cost, high-adoption LiDAR. It involves the following choices: Mature EELs, a wavelength range of 850 nm to 905 nm, scanning by MEMS mirrors, and the use of APDs as photodetectors. What do all these choices have in common? They are all mature, well-established technologies. These technologies, combined with the ToF method, could easily push LiDAR manufacturing to millions of units, which would further drive cost down. An example of MEMS LiDAR that follows the proposed path for high adoption and low cost has already been prototyped [71] (see Fig.16). Its architecture is neither strictly monostatic or bistatic, but instead hybrid: consisting of two MEMS Mirrors—one for the transmitting path and one for the receiving path. These mirrors operate as a synchronized pair, hence the name Synchronized MEMS Pair LiDAR (SyMPL). This architecture significantly simplifies the coaxial design and increases its efficiency. SyMPL operates at wavelength of 905nm, leveraging the maturity of EEL technology while benefiting from reduced susceptibility to adverse weather conditions. It is compact (Fig. 16a) and well-suited for further miniaturization, with the potential to be scaled down to the size of a matchbox. The power consumption of SyMPL is approximately 1W. Designed as a mid-range LiDAR, it has a target range of up to 25 m and FOV of 30°.

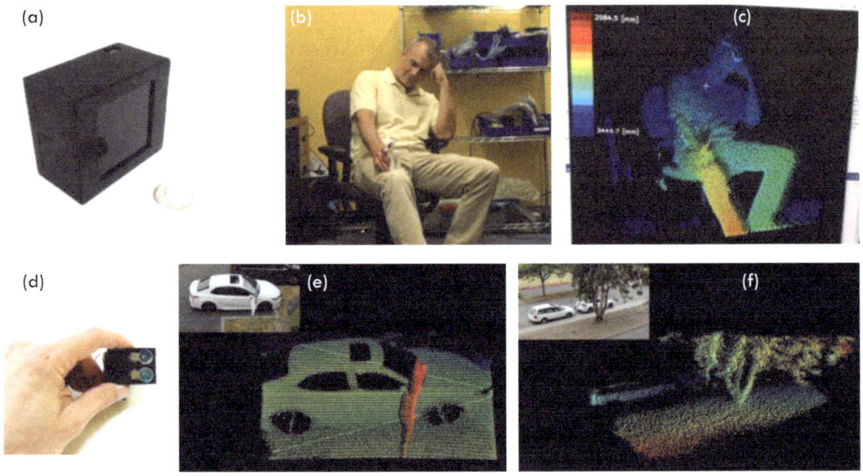

Fig. 16. a) MEMS LiDAR branded as SyMPL, b) "The Thinker" scene, c) SyMPL point cloud data from "The Thinker" scene after removal of background data, d) OEM version of SyMPL MEMS Lidar without housing weighing 49g, e) point cloud of parked car and street sign at ~15m, f) cars and tree at 20m.

We are first to acknowledge that low-cost LiDAR will not solve all problems, and there will always be room for alternative and additional fusion solutions to coexist. However, this approach could be compelling for many applications and it may indeed pave the way for the widespread adoption of LiDAR *en masse.*

3.8 Data Fusion

Data fusion, often called sensor fusion, refers to the simultaneous processing of data from multiple sensors to generate synthesized information of greater intelligence than what individual sensors can provide. Typically, advanced algorithms and special filtering techniques, such as Quaternion Kalman Filtering, are used to produce more intelligent results. A good example of sensor fusion is the integration of signals from an accelerometer, gyroscope, and magnetometer in a smart telephone [77]. By processing these signals together, sensor fusion compensates for the limitation of individual sensors, resulting in more reliable and high-quality information.

The concept of data fusion could also be applied in the optoelectronics space. A promising application is the fusion of LiDAR 3D environmental sensing capability with VGLP technology [62], aimed at enhancing performance and improving safety in application. Fig.17 illustrates the high level concept of this fused solution, referred to as PLidar. As shown in Fig. 17a, PLidar integrates LiDAR with the VGLP projector (Playzer) to enable low-latency, animated visual messaging. In this scenario, as a robot moves, the LiDAR component scans its surroundings, collects data, and continuously updates the system. Simultaneously, the VGLP projector displays real-time information, such as the robot's angle and distance from a wall (Fig.17b). The AI model processes LiDAR data and dynamically streams messages, ensuring reliable and real-time updates. The tight interaction between sensing and display significantly enhances workplace reliability and safety. Similarly, as depicted in Fig.17c, a humanoid robot can leverage data fusion to scan its surroundings using LiDAR and project tailored messages onto the ground, facilitating enhanced robot-human interaction. Additionally, by utilizing facial recognition based on LiDAR data, the system can identify individuals, further expanding its potential applications.

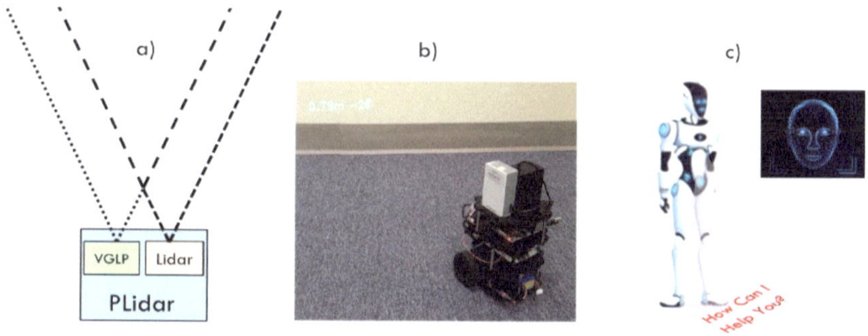

Fig. 17. a) Fusion of VGLP capability with Lidar sensing capability creates PLidar, b) PLidar in action assembled on the mini robot, c) PLidar enhancing humanoid robot capabilities.

One can envision the use of this technology in various other applications. For example, in the automotive space, a system integrating radar with VGLP projection could enhance pedestrian safety. In this scenario, the radar detects pedestrians around the vehicle, while the VGLP projector displays visual warnings directly onto the ground, alerting them not to cross the vehicle's path. This real-time interaction between sensing and display improves pedestrian-vehicle communication, ultimately enhancing safety and reducing accident risks.

4 Can MEMS Mirrors Reach One Billion Units?

It is evident from the discussion in this paper that MEMS Mirror applications are numerous (Fig.18), and we are not even scratching the surface. Many of the outlined applications are expected to generate billions of dollars. We will mention just a few to illustrate the point. MR devices, which are predicted to capture market share from smart phones, are expected to grow to hundreds of millions of units pushing this market segment to billions of dollars. The same applies to optical switching in data centers, driven by the insatiable demands from AI. Additionally, autonomous Robo-taxis, with numerous LiDARs distributed around the vehicle, represent another market opportunity of billions of dollars. On top of this, many new applications will emerge - some of which we cannot yet foresee. Taking all of this into account, one can estimate that market opportunities for MEMS Mirrors will reach tens of billions of dollars in the next decade.

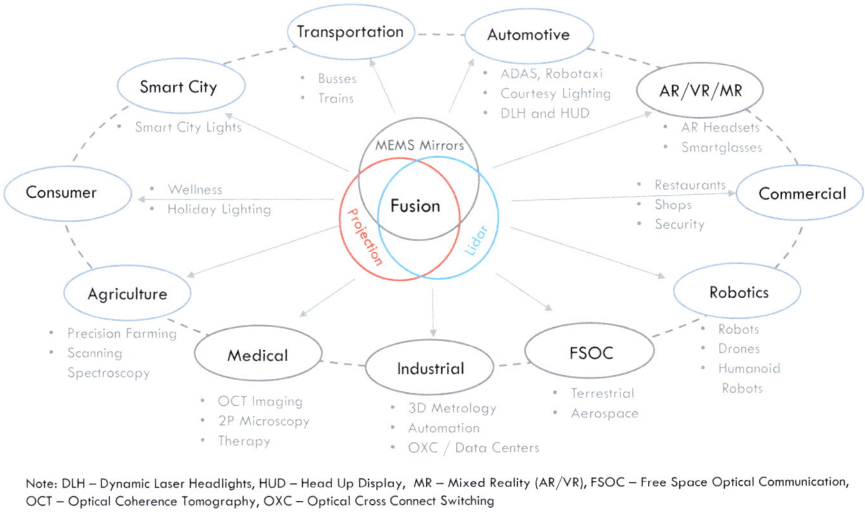

Fig. 18. MEMS Mirrors applications are numerous, and market opportunities will amount to billions of dollars.

Status of MEMS Mirrors technology. After two decades of development, MEMS mirrors are entering a phase of maturity. These devices are very reliable since they are made out of silicon crystal. Extensive testing and data have shown the ability to operate over a billions of cycles without failure. When we consider the numerous applications, some of them described here, it becomes clear that the conditions are right for exponential growth in the next decade and that the path to reaching a volume of billion units annually is entirely feasible. The confidence level is further reinforced by the fact that multiple groups of MEMS products have already followed a similar trajectory (Fig.19).

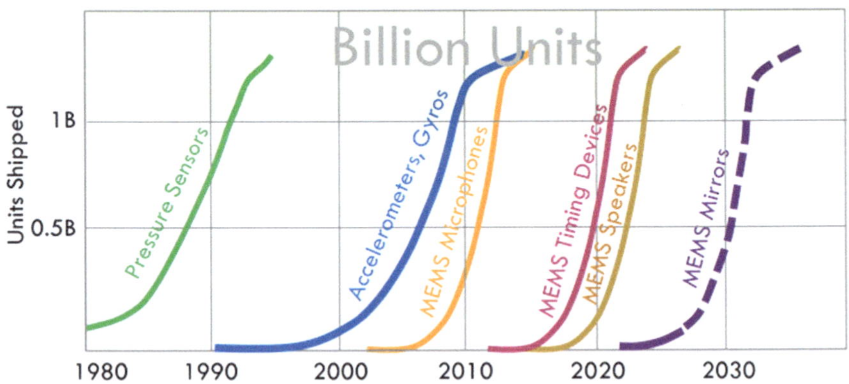

Fig. 19. MEMS mirrors – the next big wave in MEMS technology. Products from pressure sensors to speakers have already reached shipments of more than billion units annually.

5 Summary

We have reviewed MEMS technology, tracking its multiple successes that have led to its emergence as a mainstream solution for delivering reliable products in high volumes. Several product categories have already reached production volumes in the billions, attesting to the robustness and credibility of this technology.

Our analysis of MEMS mirror technology underscores the maturity of these devices, which are now on the cusp of entering a high-growth phase and are poised to replicate the success of their MEMS predecessors. Additionally, we examined the current state of LiDAR technology and proposed a potential path forward for its widespread adoption centered on MEMS LiDAR architecture.

Based on our overall analysis, we concluded that MEMS mirrors are well positioned to reach annual production volume of one billion units in the next decade, with market opportunities valued in the tens of billions of dollars.

References

1. Petersen, K.E.: Silicon as a mechanical material, Proceedings of the IEEE, 70 (5), 420-457, May (1982); doi: 10.1109/PROC.1982.12331.
2. Howe, R.T. and Muller, R.S.: Polycrystalline silicon micromechanical beams, Journal of the Electrochemical Society, 130, pp. 1420-1423 (1983).
3. Ristic, Lj., Hughes, H., and Shemansky, F.: Bulk Micromachining Technology, ch.3. In: Ristic, Lj., (ed.), Sensor Technology and Devices, Artech House, Boston (1994).
4. Ristic, Lj., Shemansky, F., Kniffin, M.L., and Hughes, H.: Surface Micromachining Technology, ch.4, In: Lj. Ristic (ed.), Sensor Technology and Devices, Artech House, Boston (1994).
5. Ristic, Lj.: CMOS technology: a base for micromachining, Microelectronics Journal, 20, (1–2), 153-169 (1989), ISSN 1879-2391; https://doi.org/10.1016/0026-2692(89)90128-6
6. Tang, W.C., Nguyen, T.-C. H., and Howe, R.T.: Laterally driven polysilicon resonant microstructures," Sensors and Actuators, 20, pp. 25-32 (1989).
7. Patent US5146389, Differential capacitor structure and method, 1992-09-08.
8. Ristic, Lj. et al.: A capacitive type accelerometer with self-test feature based on double-pinned polysilicon structure, 7th Int. Conf. SSSA, Transducers '93, Yokohama, June (1993).
9. Patent US5181156, Micromachined capacitor structure and method for making, 1993-01-19
10. Patent US5545912, Electronic device enclosure including a conductive cap and substrate, 1996-08-1.
11. Vujosevic, M., Shah, M.: Stress isolation of an accelerometer die in a post-molded plastic package--The impact of die coat coverage. In: SAE Technical Paper Series 1999, SAE International Congress and Exposition, Detroit, Michigan, March 1-4 (1999).
12. Patent US5632854, Pressure sensor method of fabrication, 1997-05-27

13. Smith, S. C.: Piezoresistance Effect in Germanium and Silicon, Phys. Rev., 94, p.42, (1954)
14. Kasten, K. et al.: Capacitive pressure sensor with monolithically integrated CMOS readout for high temperature applications, Sensors and Actuators A: Physical, 97-98, pp.83-87 (2002)
15. Stoler, A. I.: The Etching of Deep, Vertical-Walled Patterns in Silicon, RCA Rev., 31, p.271 (1970)
16. Wagner, H. A.: Electrochemically Controlled Thinning of Silicon, Bell. Syst. Tech. J.50, p.473 (1970)
17. Bean, K. D.: Anisotroping Etching of Silicon, IEEE Trans. El. Dev., ED-25, 1185 (1978)
18. X-ducerTM Sensor Element, in Motorola, Sensor Device Data Book, DL 200/D, Rev.5, 01/2003, Motorola Pressure Sensors, p.3-8 (2003)
19. Hertz, J.: STMicroelectronics Debuts 'First' Waterproof MEMS Pressure Sensor, All About Circuits, June 5 (2023); https://allaboutcircuits.com/news/st-debuts-first-waterproof-mems-pressure-sensor/
20. Gogoi, B., Vujosevic, M., and S. Petrovic,: Challenges in MEMS Packaging, Invited paper, SMTA International Congress and Exposition, Chicago, Illinois, 24-38 September (2000).
21. Frank, R., Kniffin, M.L., Ristic, Lj.: Packaging for Sensors, ch.6, p.233, in Ristic, Lj. (ed.), Sensor Technology and Devices, Artech House, Boston (1994).
22. Bever, T., Kandler, M.: Solutions for Tire Pressure Monitoring Systems, in Valldorf, J., Gessner, W. (eds.) Advanced Microsystems for Automotive Applications, pp. 261-269, Publisher Name: Springer, Berlin, Heidelberg (2003); doi.org/10.1007/978-3-540-76988-0_21
23. Freescale introduces world's smallest integrated tire pressure monitoring system, Green Car Congress, 20 Oct. (2014); https://www.greencarcongress.com/2014/10/20141020-freescale.html
24. Adorno, S., Cerini, F., Vercesi, F. (2022). Microphones. In: Vigna, B., Ferrari, P., Villa, F.F., Lasalandra, E., Zerbini, S. (eds), Silicon Sensors and Actuators. Springer, Cham. https://doi.org/10.1007/978-3-030-80135-9_15.
25. Inoue, T., Udhida, Y., Ishimoto, K. and Horimoto, Y.: Study and development of low-noise MEMS acoustic sensors. OMRON TECHNICS Vol.50.012EN (2019).
26. Wenzel, T, and Retting, R.: Design of MEMS Microphone Protective Membranes for continuous outdoor applications. Applied Acoustics, 183, December (2021), https://doi.org/10.1016/j.apacoust.2021.108304.
27. Wasisto, H.S., Anzinger, S., Acanfora, G. et al.: Acoustically semitransparent nanofibrous meshes appraised by high signal-to-noise-ratio MEMS microphones. Commun Eng 3, 136 (2024). https://doi.org/10.1038/s44172-024-00283-4.
28. Patent CN112492489B, Waterproof MEMS microphone, 2022-06-24.
29. Seo, K.; Park, J.; Kim, H.; Kim, D.; Ur, S.; Yi, S.: Micromachined piezoelectric microspeakers fabricated with high quality AlN thin film, Integr. Ferroelectr. 95, 74–82 (2007)
30. Sturtzer, E., Shahosseini, I., Pillonnet, G., Lefeuvre, E,, Lemarquand, G.: High fidelity microelectromechanical system electrodynamic micro-speaker characterization. J. Appl. Phys. 113, 214905 (2013).
31. Garud, M.V., Pratap, R.: A novel MEMS speaker with peripheral electrostatic actuation. J. Microelectromechanical Syst., 29, 592–599 (2020).
32. Fei, W., Zhou, J., Guo, W.: Low-voltage driven graphene foam thermoacoustic speaker. Small 2015, 11, 2252–2256 (2015).
33. Gazzola, C., Zega, V., Cerini,F., Adorno, S., and Corigliano, A.: On the Design and Modeling of a Full-Range Piezoelectric MEMS Loudspeaker for In-Ear Applications. J. of Microelectromechanical Systems, 32, (6), 626-637, Dec. (2023); doi: 10.1109/JMEMS.2023.3312254.

34. Van Beek, J.T.M., and Puers, R.: A review of MEMS oscillators for frequency reference and timing applications. J. Micromechanics and Microengineering, 22 (1), 013001 (2012); doi: 10.1088/0960-1317/22/1/013001
35. Chen, C.Y., Li, M.H., Li, S.S.: CMOS-MEMS Resonators and Oscillators: A Review, Sensors and Materials, 30 (4), 733-756 (2018)
36. Lutz, M., Partridge, Tabatabaei, A.S., and Sevalia, P.: Programmable MEMS-based silicon timing solutions change the game. NSTI-Nanotech 2011, www.nsti.org, ISBN 978-1-4398-7139-3, vol.2, (2011).
37. Partridge, A., and McDonald, J.: MEMS to replace quartz oscillators as frequency sources. NASA Tech Brief, vol.30, (6), (2006).
38. Milanović, V., Matus., A.G., McCormick, D.T., Gimble-Less Monolithic Silicon Actuators for Tip-Tilt-Piston Micromirror Applications, IEEE J. Sel. Topics in Quantum Electronics, 10 (3), 462-471 (2004).
39. Milanović, V., Multilevel-beam SOI–MEMS fabrication and applications, J. Microelectromech. Syst., 13 (1), 19–30, Feb. (2004).
40. Ristic, Lj.: MEMS mirrors: The next big wave in MEMS technology, Laser Focus World, July, (2023), https://www.laserfocusworld.com/optics/article/14296740/mems-mirrors-the-next-big-wave-in-mems-technology
41. https://www.mirrorcletech.com/wp/products/mems-mirrors/
42. https://www.mirrorcletech.com/wp/products/system-solutions/playzer/
43. Kasturi, A., Milanović, V., Hu, F., Kim, H. J., Ho, D., Lovell, D.: MEMS Mirror Module for Programmable Light System, SPIE 2019 OPTO Conf., San Francisco, CA. February 3, (2019)
44. Azuma, R.: A Survey of Augmented Reality. In Presence: Teleoperators and Virtual Environments, 6 (4), 355-385, Aug. (1997).
45. Azuma,R. Hoff, B., Neely, H., and Safarty, R.: A Motion-Stabilized Outdoor Augmented Reality System. In: Proc. IEEE Virtual Reality '99. Houston, 252-259, TX, 13-17 Mar. (1999).
46. Smartglasses Light Drive | Bosch Sensortec; https://www.bosch-sensortec.com/products/display-solutions/smartglasses-light-drive/
47. Leitgeb, E., Gebhart, M., and Birnbacher, U.: Optical Networks, Last Mile Access and Applications, Journal of Optical Fiber Commun. Res., 2(1), 56–85, (2005).
48. Chaleshtory, Z.N., Gholami, A., Ghassemlooy, Z., Sedghi M.: Experimental Investigation of Environment Effects on the FSO Link with Turbulence. IEEE Photon Technol Lett (2017) 29:1435–8. doi:10.1109/LPT.2017.2723569.
49. Hong, Y-Q., Kwon, D-H., Choi, J-Y., Ha, I-H, Shin, W-H, Han, S-K.: SOA-based Multilevel Polarization Shift On-Off Keying Transmission for Free-Space Optical Communication. Photonics (2021) 8:100–8. doi:10.3390/photonics8040100.
50. Kingsbury, R. W. et al.: Fast-Steering Solutions for Cubesat-Scale Optical Communications, ICSO 2014, Tenerife, Canary Islands, Spain, 7 – 10 Oct. 2014; https://dspace.mit.edu/bitstream/handle/1721.1/132241/105630G.pdf?sequence=2&isAllowed=y
51. Hammel, F. D. Eaton and S. L. Lachinova.: Atmospheric Channel Effects on Free-Space Laser Communication, Journal of Optical Fiber Communications, vol. 3(2), 111–158, (2006).
52. Chauhan, S., Miglani, R., Kansal, L., Gaba, G.S., Masud M. Performance Analysis and Enhancement of Free Space Optical Links for Developing State-Of-TheArt Smart City Framework. Photonics (2020) 7:132; doi:10.3390/ photonics7040132.

53. A pioneering launch – compact satellite PIXL-1 carries the world's smallest laser terminal into orbit; https://www.dlr.de/en/latest/news/2021/01/20210124_pioneering-launch-compact-satellite-with-smallest-laser-terminal
54. Taara - A Google X Moonshot; https://x.company/projects/taara/
55. https://taaraprojects.com
56. Bishop, D.J., C. R. Giles, and G. P. Austin,: The Lucent Lambda-Router: MEMS technology of the future here today, IEEE Comm. Mag., 40, (3), 75–79 (2002)
57. Yamaguchi, J. et al.: High-yield fabrication methods for MEMS tilt mirror array for optical switches, NTT Technical Review, 5 (10), Oct. (2007).
58. Urata, R. et al., Mission Apollo: Landing optical circuit switching at datacenter scale, Aug. (2022), https://doi.org./10.48550/arXiv.2208.10041
59. Milanović, V., and Ristic, Lj.: MEMS Mirror Array Optical Switching Technology by Advanced Packaging, Project proposal for U.S. CHIPS and Science Act funding, 2025-NIST-CHIPS-NAPMP-01, December (2024).
60. Deng, P., Kane, T.: Alharbi, O., Reconfigurable free space optical data center network using gimbal-less MEMS retroreflective acquisition and tracking, Proc. of SPIE Vol. 10524 1052403-1, (2018); doi: 10.1117/12.2295831
61. Palinko, O. et al.: Intention Indication for Human Aware Robot Navigation, doi.org/10.5220/0009167900640074
62. Lovell, D. et al.: Optical MEMS Enable Next Generation Solutions for Robot Vision and Human-Robot Interaction, Proceedings of the SPIE, vol.12013, MOEMS and Miniaturized Systems XXI, 1201304, March (2022); https:// doi.org/10.1117/12.2613458
63. Li, J., Ibanez-Guzman, J.: Lidar for Autonomous Driving: The principles, challenges, and trends for automotive lidar and perception systems, Apr. 2020, arXiv:2004.08467v1 [cs.RO] 17 Apr (2020).
64. Boucart, J.: Shining a Spotlight on Laser Sources for Automotive Exterior LiDAR, e-motec.net (E-Mobility Technology International), Sept. 15 (2021); https://www.e-motec.net/laser-sources-for-automotive/
65. Michaud, S., J.-F. Lalonde, and P. Gigure, "Towards characterizing the behavior of LiDARs in snowy conditions," in IEEE/RSJ IROS Workshop on Planning, Perception and Navigation for Intelligent Vehicles, (2015).
66. Kutila, M., Pyykonen, P., Holzhuter,H., Colomb, M., and Duthon, P.: "Automotive LiDAR performance verification in fog and rain," in 21st IEEE International Conference on Intelligent Transportation Systems (ITSC), (2018).
67. Sun, X.: Review of photodetectors for space Lidars, Sensors 2024, 24, 6620 (2024); doi.org/10.3390/s24206620
68. Li, L.: Time of Flight Camera – An Introduction, Technical White Paper, SLOA190B – January 2014, Revised May (2014)
69. Gao, S.: Hui, R.: Frequency-modulated continuous-wave lidar using I/Q modulator for simplified heterodyne detection, Optics Letters, 37 (11), 2022-2024, (2012); https://doi.org/10.1364/OL.37.002022
70. Raj, T. et al.: A survey on Lidar scanning mechanism, Electronics 2020, 9, 741 (2020); doi.org/10.3390/electronics9050741
71. Kasturi, A. et al.: Comparison of MEMS mirror LiDAR architectures", Proc. SPIE 11293, MOEMS and Miniaturized Systems XIX, 112930B, February (2020), https://doi.org/10.1117/12.2556248
72. Amzajerdian, et al.: "Imaging Flash Lidar for Autonomous Safe Landing and Spacecraft Proximity Operation," in AIAA SPACE, Long Beach, U.S.A, (2016).

73. Mcmanamon, P.F., et al., "Optical phased array technology," Proceedings of the IEEE, 84, 268–298 (1996)
74. Acoleyen, V. et al.: Integrated Optical Beam Steerers. In Proceedings 2013 Optical Fiber Communication Conference and Exposition and the National Fiber Optic Engineers Conference (OFC/NFOEC), Anaheim, CA, USA, pp. 6–8, 17–21 March (2013)
75. Argueta, V.: Exploring LIDAR Lens Design: Innovations and Applications in Optical Engineering, OFH, Sept.16 (2024); https://www.opticsforhire.com/blog/lidar-lens-design-optical-layout/
76. Lee, X., Wang, C.: Optical design for uniform scanning in MEMS-based 3D imaging lidar, Applied Optics, 54 (9), 2219-2223 (2015); doi.org./10.1364/AO.54.002219
77. Ristic, Lj.: Sensor fusion and MEMS technology for 10-DoF solutions, EE Times, Sept. 3, (2012).

Subwavelength Spatial Light Modulator powering the Next-Generation of Smart Eyewear

Andrzej Kaczorowski[1[0000-0001-5949-8200]], Aaron Demolder[1[0000-0002-4394-5915]], Gebirie Yizengaw Belay[1[0000-0003-1470-3543]], Edward Buckley[1], Theodore Marescaux[1], Dmitri Choutov[1] and Michael Noonen[1]

[1] Swave Photonics, Leuven, Belgium
ak@swave.io

Abstract. Computer-Generated Holography (CGH) is an emerging display technology that offers unique capabilities, such as reconstructing 3D objects and correcting optical aberrations (including those of the human eye) in software.
The capabilities of this technology were known since the 1990s thanks to the pioneering work at the Massachusetts Institute of Technology. Despite a number of commercialization attempts, the potential of this technology has not been fully realized in a consumer product. Within the last decade, the field saw renewed interest thanks to its applications in Augmented Reality (AR) and smart eyewear. The advantages of CGH address numerous issues that Augmented Reality is currently facing.
The two bottlenecks that ought to be overcome while introducing a CGH-powered smart eyewear to the market are the lack of sufficiently advanced light modulation technology and high-fidelity hologram generation algorithms that run in real time.
Swave Photonics addresses those two challenges by introducing a Holographic Extended Reality (HXR) chipset consisting of a Subwavelength Spatial Light Modulator and its compute companion chip - the Holographic Processing Unit.

Keywords: Smart Eyewear, Augmented Reality, Computer-Generated Holography, Holographic Display, Display Technology, 3D Displays.

1 Introduction

1.1 Computer-Generated Holography – History and Modulation Technology

The principles of Computer-Generated Holography (CGH) have been well known for several decades. Most notably, as far back as the 1980s, the Massachusetts Institute of Technology conducted research into Holographic Displays based on Acousto-Optic Modulators (termed Mark I display) together with a custom-designed supercomputer, a Connection Machine 2 (CM-2) to perform the computation [1]. The 1D nature of Acousto-Optic Modulators meant a complex scanning arrangement in order to get a sufficient Field of View and viewing window combination.

S. Kurimura and T. Yagi (eds.), *Proceedings of Laser Display and Lighting Conference 2025*, Springer Proceedings in Physics 429,
https://doi.org/10.1007/978-981-96-9782-3_18

As the Liquid-Crystal-on-Silicon (LCoS) devices became available and began scaling up their resolution while scaling down the pixel pitch, the holography researchers used their capability to display complex CGH patterns. The current state-of-the art in LCoS is the pixel pitch of 3µm with a HD resolution, some devices go further into 4K. Reducing the pixel pitch significantly below this figure has been a yet-unaddressed challenge due to the range of negative effects that get stronger as the pixel size is shrunk, like the Fringe Field Effect.

1.2 Augmented Reality smart eyewear based on Computer-Generated Holography

Fundamentally, there are two ways to implement Augmented Reality (AR) devices using Computer-Generated Holography. The difference between them is the region of the image formation between the modulator and the image it displays. In one example, the images are formed in the Fresnel region behind the surface of the chip. While in the second, the images appear in close proximity to the surface of the modulator (in the near field). The choice of the diffraction region determines the optical configuration that needs to be used.

In our work, we will focus on the configuration where the chip is re-imaged onto the eye pupil and becomes the eyebox - which is the window through which the field of view (FoV) is visible - of the device. This configuration is presented in Fig. 1. It is easy to see that the diffraction angle of the device becomes the Field of View (FoV), while the physical size of it becomes the eyebox.

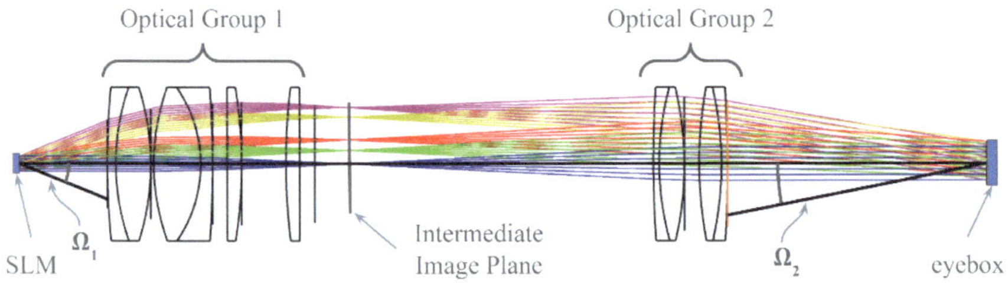

Fig. 1. A 4f optical configuration of a holographic display, allowing to trade off Field of View (diffraction angle) for the eyebox size. Note that in this illustration, only a portion of the diffraction angle coming from the HXR chip (Ω_1) is used as an input to the optics.

1.3 Etendue conservation in a holographic display

It is possible to trade-off FoV for the eyebox, but the product of the two (referred to as etendue) has to remain constant. Given area of the modulator A and the angle it emanates Ω, the etendue conservation can be expressed in Eq. (1).

$$A \times \Omega = \text{eyebox} \times \text{FoV} = constant \tag{1}$$

Let us take a hypothetical LCoS device with 4μm pixel pitch and 2K resolution as an example. The pixel pitch of the device dictates its diffraction angle of 7.5 degrees and the size of the panel is dictated by the number of pixels and their pitch - 8.2mm.

Using a simple 4f system (such as the one shown in Fig. 1), it is possible to trade-off the eyebox for the Field of View; this is an important consideration in augmented reality (AR) because a large eyebox is a key requirement to accommodate for the large range of inter-pupillary distances (IPDs) of any viewer which is a fundamental requirement of mass-market AR.

Using this hypothetical device and trading off FoV and the eyebox, the possible configurations are 30-degree FoV with a 2mm eyebox or a 10mm eyebox with an FoV of 6 degrees. In fact, achieving a sufficiently big FoV and eyebox simultaneously is the biggest problem the industry is facing, which maps simply to the number and density of pixels of the modulator.

2 Requirements for a consumer-acceptable smart eyewear

It is widely agreed within the technology industry that Augmented Reality and smart eyewear is the future platform that is going to replace mobile phones in the realm of human-machine interaction.

However, a number of constraints imposed on those wearables are very often conflicting with one another. The slim form factor required for the glasses to be all-day-wearable is in direct opposition to the wide field of view desired for full immersion into the displayed 3D content. The requirement to have a bright display that can be used outdoors is competing with the fact that the device should be all-day usable given a reasonable capacity battery. So far, none of the existing technologies have been able to meet those requirements to a degree to make a successful consumer product.

To make matters worse, the fact that the smart eyewear merges the virtual content with the physical environment, implies that the virtual overlay is meant to appear at the same depth as the objects it is augmenting. If that is not the case, the glasses risk triggering a condition known as Vergence-Accommodation Conflict (VAC) which appears when accommodation (the distance where the user focuses) is not coherent with vergence (the rotation of the eyeball).

Finally, over 60% of the adult population globally wear prescription eyewear, while over 40% require advanced correction [2], such as cylindrical lenses to counteract astigmatism. A pair of glasses is today, primarily, a medical device to correct vision. Smart eyewear must accommodate for the needs of those who require a prescription correction, and for those who do not, with as little complexity as possible. Augmenting one's reality and enabling interaction with technology should not significantly obstruct nor degrade the ordinary view of the world.

3 Holographic eXtended Reality (HXR) solution proposed by Swave Photonics

Swave photonics has developed a proprietary light modulation technology with a sub-wavelength pixel pitch. The modulator, based on non-volatile phase change materials on silicon, is fabricated using 90%+ standard CMOS technology, ensuring scalability and cost-effectiveness. The pixel pitch of <300nm ensures a large diffraction angle and the ability to pack a large number of those nano-pixels into a small area (this is illustrated in Fig. 2). A 16Kx16K modulator is little over 5mm x 5mm in size. This extraordinary amount of etendue means that it is possible to simultaneously achieve a large eyebox and Field of View without the necessity for a pupil-replicating waveguide or other exotic optical elements.

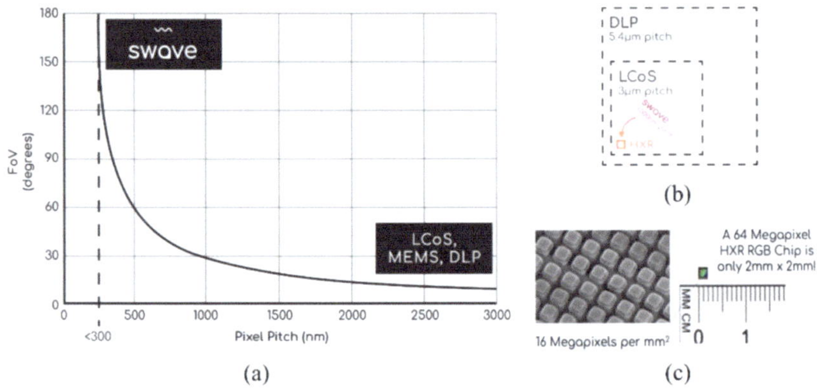

Fig. 2. The properties of the HXR modulator and its effect on the optical system (a) The Field of View (diffraction angle) as a function of the pixel pitch, (b) HXR pixel to scale with other technologies, (c) Pixel density of the HXR chip illustrated.

A display based on Computer-Generated Holography belongs to the category of Computational Displays. It means that both the modulator, as well as the algorithm computing the holographic patterns displayed on it are necessary for the operation of the display. Computing holograms has historically been a challenging task. To illustrate the magnitude of the computational challenge, it is worth pointing out that MIT had commissioned a build of a supercomputer to handle running of the Mark I display in 1990. In order to address this, Swave Photonics is developing the chipset composed of a Spatial Light Modulator as well as the companion compute accelerator, called the Holographic Processing Unit (HPU).

Implementing holography in wearable devices is a non-trivial task. Once it is done, however, there are numerous benefits of holographic projection that one enjoys at no extra cost. First and foremost, holography is a wavefront control technology. As a result, it naturally allows for smooth and precise control of focusing distance. Taking this one step further, it is also possible to encode aberration correction into the wavefront.

This capability can be used to correct imperfections of optical systems (for instance in order to enable further miniaturization and manufacturing cost reduction) [3]. Of course, the one optical system that humans deal with on a daily basis is our own eye, and optical imperfections of those can also be easily accounted for and corrected. That includes a focus error, astigmatism and further errors (such as spherical aberration, coma, trefoil and higher-order aberrations).

Another benefit of holography which has been overlooked so far is the fact that it is inherently a beam-steering technology. As a result, the sparser the image is, the brighter it will appear, since all the available light is being re-directed to pixels of the image that are "on". It has been previously demonstrated that this leads to very bright images that can be formed with modest amounts of laser light [4].

3.1 Results

While the dynamic modulator has been in development, Swave produced a number of static modulators with holographic patterns imprinted on them. The aim has been to demonstrate the potential of this technology, test the assumptions about optical systems and de-risk algorithmic optimizations. Fig. 3 shows a picture of a 12-inch Silicon wafer filled with holographic chips. Those chips have then been diced and integrated into optical systems. The result of this can be seen in Fig 4.

Fig. 3. Example 12-inch silicon wafer with a multitude of static RGB holographic patterns encoded into it.

Just as search engines have been an integral part of the success of the early Personal Computer, and current mobile devices - Visual Search (the ability to search for objects or information via computer vision) is a unique function and the key benefit given by the adoption of Smart Glasses.

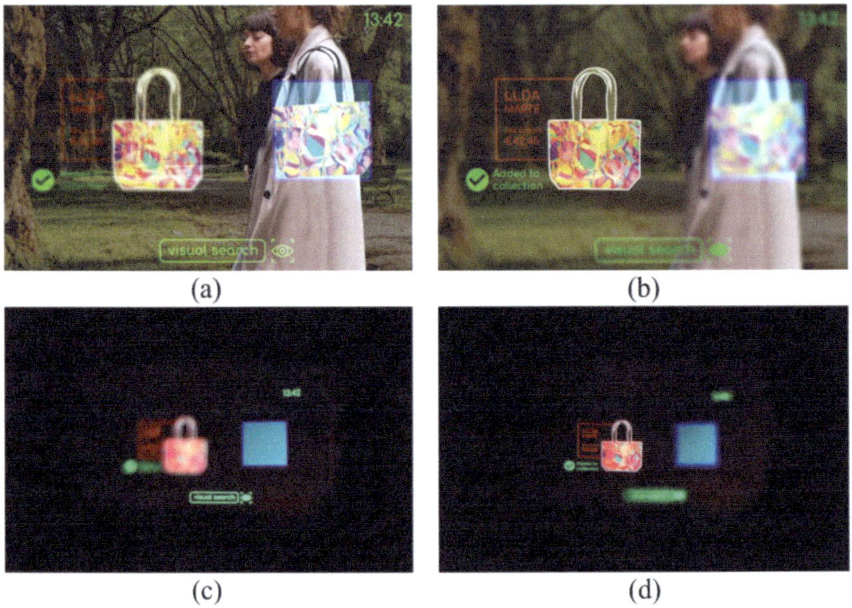

Fig. 4. Illustration of Visual Search: (a) The desired product is being highlighted by the display, (b) The purchase suggestions are displayed in the user's personal space, (c)-(d) The holographically-projected image captured from a holographic device at (a) 5m and (b) 1m.

This is the use case scenario depicted in Fig. 4, where an object of interest is being searched and displayed in one's environment. Since, as mentioned previously, it is crucial to show realistic depth perception to correctly augment our environment, we demonstrate two planes of projection, which focus at 5m (Fig 4c) and 1m (Fig. 4d).

4 Conclusions

Augmented Reality smart eyewear is considered to be the platform that will replace smartphones and redefine how humans interact with technology. Despite the substantial efforts from the industry, there isn't yet a successful mass consumer product. This is because the requirements of AR are inherently very challenging with numerous opposing constraints.

To address those issues, Swave introduces the Holographic eXtended Reality (HXR) chipset. The Swave Spatial Light Modulator with a subwavelength pixel pitch simultaneously provides a wide diffraction angle, and unprecedented pixel density. This capability opens a path towards a viable smart eyewear with a large Field of View and eyebox without the necessity to use pupil-replicating waveguides. The modulator chip is supported by its compute companion – the Holographic Processing Unit, which computes high-quality holographic patterns in real time and at a power consumption level compatible with smart eyewear.

This innovation enables a truly-holographic pair of smart eyewear which inherits all the benefits of computer-generated holographic display, namely, the capability to dynamically change focal depth, prescription correction and daylight brightness due to laser illumination.

References

1. St.-Hilaire, P., Benton, S.A., Lucente, M., Hubel, P.M. Color images with the MIT holographic video display. In: Practical Holography VI, Proc. SPIE 1667 (1992). DOI: 10.1117/12.59642
2. Hashemi, H., Fotouhi, A., Yekta, A., Pakzad, R., Ostadimoghaddam, H., Khabazkhoob, M. (2018). Global and regional estimates of prevalence of refractive errors: Systematic review and meta-analysis. Journal of Current Ophthalmology, 30(1), 3–22. DOI: 10.1016/j.joco.2017.08.009
3. Kaczorowski, A. (2018) Adaptive Aberration Correction for Holographic Projectors. Doctoral Thesis, Apollo - University of Cambridge Repository. DOI: 10.17863/CAM.17173
4. Newman, A. J., Widjanarko, T., Spiess, A. O., Milne, D. F., Wilkinson, T. D., Kaczorowski, A. (2021). An intrinsically eye safe approach to high apparent brightness augmented reality displays using computer-generated holography. Journal of the Society for Information Display, 29(11), 840-851. DOI: 10.1002/jsid.1082

Part IV
Imaging and Lighting LDC4

Pseudo-Direct LiDAR Based on an Ultra-High-Speed Charge-Domain Compressive Sensing Image Sensor

Keiichiro Kagawa[1]

[1] Shizuoka University, Hamamatsu Shizuoka 432-8011, Japan
kagawa@idl.rie.shizuoka.ac.jp

Abstract. Time-compressive computational CMOS image sensors can realize a new implementation of light detection and ranging (LiDAR), i.e., pseudo-direct time-of-flight (pseudo-dToF) depth imaging. They are based on the multi-tap charge modulator typically used in indirect ToF (iToF) image sensors and compressive sensing of high-speed optical temporal signals in the charge domain at each pixel. Therefore, large digital circuits for time-to-digital conversion or histogram builder required in conventional dToF image sensors based on the single photon avalanche diode (SPAD) are not necessary. Nevertheless, they can detect multiple reflections under multipath interference, which is one of the advantages of dToF. Exploiting the small pixel size of the pseudo-dToF image sensor, high-resolution image sensors will be possible. Experimental results for a dual-reflection scene demonstrate the benefit of the pseudo-dToF.

Keywords: CMOS image sensor, LiDAR, Compressive sensing, Charge-domain signal compression.

1 Pseudo-Direct LiDAR

1.1 Conventional dToF and iToF

Recently, LiDAR image sensors[1] attract significant attention in applications such as autonomous driving, gesture recognition, and so on. There are two types of LiDAR image sensors: direct time-of-flight (dToF) and indirect ToF (iToF) image sensors. dToF image sensors are composed of single-photon avalanche diodes, time-to-digital converters (TDCs), and histogram builders. Because dToF image sensors acquire the entire waveform of the received optical signal, they are much less influenced by multipath interference (MPI)[2,3]. However, due to the huge digital circuits, pixel count is limited, or the TDCs and histogram builders are shared in columns or rows. iToF image sensors utilize multi-tap charge modulators, and no huge additional circuits are required in pixel. Therefore, it is possible to embody high spatial resolution image sensors. However, as the received optical signal is detected as correlation values between the optical signal and time-window functions applied to the charge modulators, the measured depth becomes erroneous when multipath interference exists.

S. Kurimura and T. Yagi (eds.), *Proceedings of Laser Display and Lighting Conference 2025*, Springer Proceedings in Physics 429,
https://doi.org/10.1007/978-981-96-9782-3_19

1.2 Pseudo-Direct LiDAR based on Compressive Sensing

We have proposed a pseudo-dToF image sensor based on iToF image sensor[4-5]. This image sensor was motivated by the concept of computational photography[6-7], which modulates or multiplexes three-dimensional (x, y, and t) optical signals in the charge domain, then reconstructs it by post-processing.

Because optical signals received by the image sensor in the pulse-based LiDAR are very sparse, we can compress the signal in the physical domain based on the compressive sensing[8-9]. The mathematical representation of compressive sensing is briefly mentioned below. When we consider an N-dimensional column vector x, an $M \times N$ matrix $\boldsymbol{A}$, and an M-dimensional column vector $\boldsymbol{y}$, the relationship is denoted by

$$\boldsymbol{y} = \boldsymbol{A}\boldsymbol{x}. \tag{1}$$

Note that $\boldsymbol{x}$ is the original temporal light intensity signal, and $\boldsymbol{y}$ is the signal measured through a known measurement matrix $\boldsymbol{A}$. When $N > M$, the original signal $\boldsymbol{x}$ is compressed. We can reproduce the original light signal $\boldsymbol{x}$ by solving the inverse problem from $\boldsymbol{y}$ and $\boldsymbol{A}$ based on a sparsity constraint. Here, sparsity means that most of $\boldsymbol{x}$ **or** $\boldsymbol{\Phi x}$ ($\boldsymbol{\Phi}$ is any transformation matrix ($N \times N$)). are zero and only limited number of the elements have non-zero values. $\boldsymbol{A}$ describes the spatial point spread function of imaging optics, time window functions for temporal coded shutters, and temporal response of the image sensor.

This image sensor temporally compresses the received optical signal pixelwise in a single shot. Therefore, it is suitable for capturing dynamic scenes. Compression requires no additional circuit in pixel because the signal is compressed with random coded shutters by the multi-tap charge modulator in the charge domain. Fig. 1 compares conventional dToF, iToF, and the proposed pseudo-dToF LiDARs.

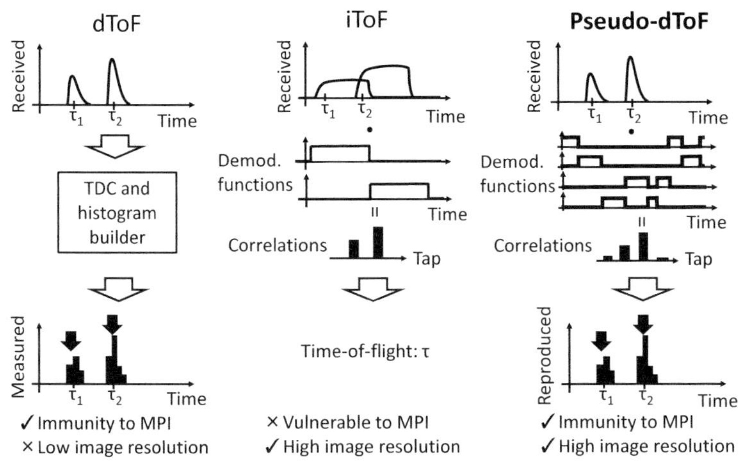

Fig. 1. Comparison of conventional dToF, iToF, and our pseudo-dToF.

2 Prototype CMOS Image Sensor

As shown in Fig. 2(a), the time-compressive computational CMOS image sensor is composed of an array of macropixels with 2×2 subpixels. Each subpixel is implemented by a four-tap charge modulator. To realize ultrafast charge modulation, the lateral electric field charge modulator (LEFM) [10-11] was adopted. The function of the four-tap charge modulator is depicted in Fig. 2(b). The temporal photocurrent signal $i_{ph}(t)$ is multiplied with a time window function or exposure code $w_i(t)$ by each multiplier implemented as a charge transfer gate and integrated in each storage area in the charge domain. In temporal compressive sensing, binary random time-window functions or shutter codes shown in Fig. 3, for example, are applied to the charge modulators repeatedly. Finally, four compressed images are obtained. The readout circuits of this sensor are the same as ordinary CMOS image sensors, i.e., pixel source followers, correlated double sampling circuits, column-parallel analog-to-digital converters, and a serializer. Table 1 summarizes the specification of a prototype pseudo-dToF CMOS image sensor.

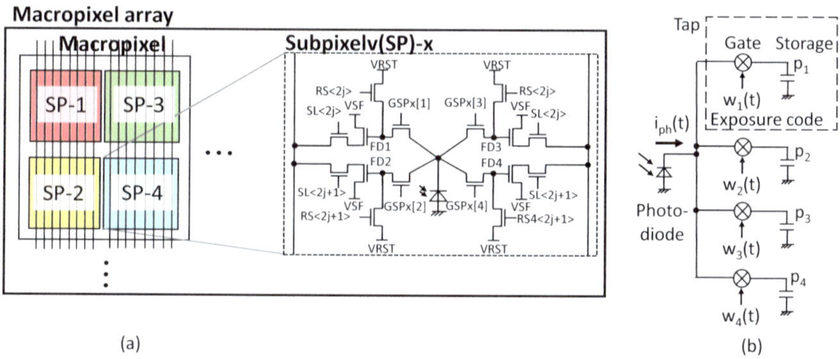

Fig. 2. (a) 2×2-pixel block structure and pixel circuit. (b) 4-tap charge modulator.

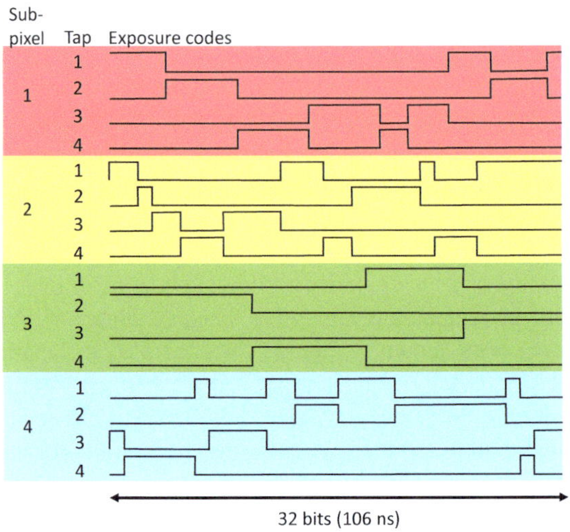

Fig. 3. An example of exposure codes.

Table 1. Chip specifications.

Parameter	Value
Subpixel size	11.2μm×11.2μm
Subpixel count	212×218
Tap count per subpixel	4
Number of shutter codes	16
Maximum shutter length	256 bits
Maximum shutter clock	832MHz
Minimum time window width	~3ns
Image readout frame rate	21fps

3 Experiments

To demonstrate the immunity to MPI of pseudo-dToF, a scene composed of two bounces was captured. To optimize the exposure codes and to reduce the signal reproduction time, the framework of deep sensing was utilized, where the exposure codes are optimized while the network is trained[12-14]. The sensor presented in [4-5] was used. The clock frequency of the exposure codes was 607MHz, and the exposure code length was 32 bits. For reproduction, the scheme in Fig. 4(a) was used. A reference board was placed at 1.6m away from the camera. Two objects, "C" and a pillar, were placed at +1.6m and +2.0m from the reference board, respectively. These objects were seen through a weak diffusive sheet placed at +0.4m. Fig. 4(b) shows that the depths of the objects "C" and the pillar were successfully obtained. These shapes are inherently blurry due to the scattering by the diffusive sheet.

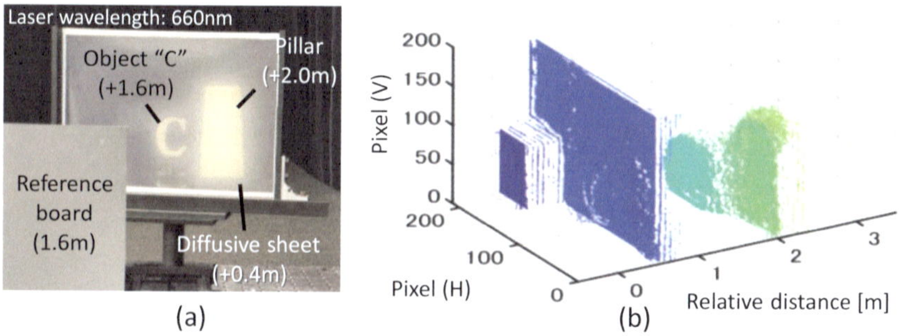

Fig. 4. (a) A scene setup. (b) The reproduced depth.

4 Conclusions

Pseudo dToF based on an iToF image sensor and compressive sensing combined with deep learning virtually implements dToF imaging and offers high spatio-temporal sampling efficiency so that high resolution LiDAR image sensors will be possible. Immunity to MPI of our system was demonstrated.

Acknowledgements

This work was partially supported by JST, CREST, JPMJCR22C1, and Grants-in-Aid for Scientific Research (A) Number 24H00313.

References

1. F. Remondino and D. Stoppa: TOF Range-Imaging Cameras. Springer (2013).
2. A. Bhandari, A. Kadambi, R. Whyte, C. Barsi, M. Feigin, A. Dorrington, and R. Raskar: Resolving multipath interference in time-of-flight imaging via modulation frequency diversity and sparse regularization, Optics Letters, vol. 39, pp. 1705–1708 (2014).
3. N. Naik, A. Kadambi, C. Rhemann, S. Izadi, R. Raskar, and S. Kang: A light transport model for mitigating multipath interference in time-of-flight sensors, In IEEE Computer Society Conference on Computer Vision and Pattern Recognition, pp. 73–81 (2015).
4. K. Kagawa, M. Horio, A. N. Pham, T. Ibrahim, S. Okihara, T. Furuhashi, T. Takasawa, K. Yasutomi, S. Kawahito, and H. Nagahara: A dual-mode 303-Megaframes-per-second charge-domain time-compressive computational CMOS image sensor, MDPI Sensors, vol. 22, Article 1953 (2022).
5. Anh Ngoc Pham, Thoriq Ibrahim, Keita Yasutomi, Shoji Kawahito, Hajime Nagahara, Keiichiro Kagawa: Depth quality improvement with a 607 MHz time-compressive computational pseudo-dToF image sensor, MDPI Sensors, Vol. 23, Issue 23, Article 9332 (2023).

6. R. Raskar, J. Tumblin, A. Mohan, A. Agrawal, and Y. Li: Computational Photography, Eurographics, Tutorial (2007).
7. H. Nagahara, T. Sonoda, K. Endo, Y. Sugiyama, and R. Taniguchi: High-speed imaging using CMOS image sensor with quasi pixel-wise exposure, Proc. IEEE International Conference on Computational Photography, pp. 81–91 (2016).
8. E. J. Cades, M. B. Wakin, "An introduction to compressive sampling," IEEE Signal Processing Mag., Vol. 25, pp. 21–30 (2008)
9. R. G. Baraniuk, "Compressive sensing," IEEE Signal Processing Mag., Vol. 24, pp. 118–121 (2007).
10. S. Kawahito, G. Baek, Z. Li, S. Han, M. Seo, K. Yasutomi, K. Kagawa: CMOS lock-in pixel image sensors with lateral electric field control for time-resolved imaging, Proc. Int'l Image Sensor Workshop, pp. 361–364 (2013).
11. K. Yasutomi, Y. Okura, K. Kagawa, S. Kawahito: A sub-100μm-range-resolution time-of-flight range image sensor with three-tap lock-in pixels, non-overlapping gate clock, and reference plane sampling, IEEE J. Solid-State Circuits, vol. 54, pp. 2291–2303 (2019).
12. M. Yoshida, A. Torii, M. Okutomi, K. Endo, Y. Sugiyama, R. Taniguchi, and H. Nagahara: Joint optimization for compressive video sensing and reconstruction under hardware constraints, European Conf. on Computer Vision, pp. 634-649 (2018).
13. Michitaka Yoshida, Daisuke Hayashi, Lioe De Xing, Keita Yasutomi, Shoji Kawahito, Keiichiro Kagawa, Hajime Nagahara: Single-shot efficient depth imaging based on time-compressive CMOS image sensor, in IEEE Sensors, Paper 6371 (2024).
14. Keiichiro Kagawa, Keita Yasutomi, Michitaka Yoshida, Daisuke Hayashi, De Xing Lioe, Shoji Kawahito, Hajime Nagahara: Pseudo-direct LiDAR by deep-learning-assisted high-speed multi-tap charge modulators, in IEEE IEDM, Paper 41-5 (2024).

Quantitative phase imaging under commercially available confocal microscopy

Naru Yoneda[1,2][0000-0001-9964-5420], Joe Sakamoto[3,4][0000-0003-2295-0659], Takumi Tomoi[5][0000-0002-9407-0426], Tomomi Nemoto[3,4,6][0000-0001-6102-1495], Yosuke Tamada[7,8,9][0000-0002-4845-4867], and Osamu Matoba[1,2][0000-0001-9767-0059]

[1] Graduate School of System Informatics, Department of System Science, Kobe University, Rokkodai 1-1, Nada, Kobe, 657-8501, Japan
[2] Center of Optical Scattering Image Science, Kobe University, Rokkodai 1-1, Nada, Kobe 657-8501, Japan
[3] Biophotonics Research Group, Exploratory Research Center on Life and Living Systems, Higashiyama 5-1, Myodaiji, Okazaki, Aichi 444-8787, Japan
[4] Division of Biophotonics, National Institute for Physiological Sciences, Higashiyama 5-1, Myodaiji, Okazaki, Aichi 444-8787, Japan
[5] Faculty of Science and Technology, Department of Applied Biological Science, Tokyo University of Science, Noda, Chiba 278-8510, Japan
[6] School of Life Science, The Graduate University for Advanced Studies (SOKENDAI), Higashiyama 5-1, Myodaiji, Okazaki, Aichi, 444-8787, Japan
[7] Faculty of Engineering, Utsunomiya University, 7-1-2 Yoto, Utsunomiya 321-8585, Japan
[8] Center for Optical Research and Education (CORE), Utsunomiya University, 7-1-2 Yoto, Utsunomiya, 321-8585, Japan
[9] Robotics, Engineering and Agriculture-Technology Laboratory (REAL), Utsunomiya University, Utsunomiya, Japan
yoneda.naru@port.kobe-u.ac.jp

Abstract. In this presentation, we introduce quantitative phase imaging under commercially available confocal microscopy. By using bright-field imaging module and the transport of intensity equation, quantitative phase distribution can be obtained without any change of optics.

Keywords: Quantitative phase imaging, Transport-of-intensity equation, Confocal microscopy

Confocal microscopy is a powerful tool that enables high-spatial-resolution observation of fluorescence-labeled biological samples. Various confocal microscope systems are commercially available from different makers, making them an essential tool for biological researchers. In recent years, there has been increasing interest in quantitative phase imaging (QPI), which provides information about the phase shifts induced by biological specimens under incident light. QPI enables label-free imaging and has potential applications in multifunctional imaging, as it allows for the extraction of information such as dry mass [1], scattering parameters [2], and molecular vibrations [3]. However, commercial confocal microscopes lack built-in QPI functionality, requiring either a dedicated QPI microscope or modifications to an existing confocal microscopy. In our previous work, we proposed a method based on the transport-of-intensity

S. Kurimura and T. Yagi (eds.), *Proceedings of Laser Display and Lighting Conference 2025*, Springer Proceedings in Physics 429,
https://doi.org/10.1007/978-981-96-9782-3_20

equation (TIE) that enables QPI on a commercial confocal microscope without any modifications to the optical system [4]. In this presentation, we will introduce the proposed method, present images obtained using this technique, and discuss noise artifacts specific to our approach along with strategies to mitigate them.

A part of the work was supported by JST CREST under Grant JPMJCR22P6 and Japan Society for the Promotion of Science (JSPS) KAKENHI (20H05886) .

References

1. S. Aknoun *et al.*, "Living cell dry mass measurement using quantitative phase imaging with quadriwave lateral shearing interferometry: an accuracy and sensitivity discussion," J. Biomed. Opt. 20(12), 126009 (2015).
2. W. J. Eldridge *et al.*, "Optical Phase Measurements of Disorder Strength Link Microstructure to Cell Stiffness," Biophys. J., 112, 692 (2017).
3. M. Tamamitsu *et al.*, "Quantitative phase imaging with molecular vibrational sensitivity," Opt. Lett. 44, 3729-3732 (2019).
4. N. Yoneda *et al.*, "Transport-of-intensity phase imaging using commercially available confocal microscope," J. Biomed. Opt. 29(11) 116002 (2024).

Application of Holography for Data Storage

Xiaodi Tan[1,2,3], Kazuo Kuroda[4], Jianying Hao[1], Shujun Zheng[1], Haiyang Song[1], Shenghui Ke[1], Yongkun Lin[1], Hongjie Liu[1], Ruixian Chen[1], Rupeng Yang[1], Dakui Lin[1,2], Xiao Lin[1,3] and Yuhong Ren[1]*

[1] Information Photonics Research Center, College of Photonic & Electronic Engineering, Fujian Normal University, Fuzhou, Fujian 350117, China
[2] Key Laboratory of Opto-Electronic Science and Technology for Medicine of Ministry of Education, Fujian Normal University, Fuzhou, Fujian 350117, China
[3] Fujian Provincial Key Laboratory of Photonics Technology, Fujian Normal University, Fuzhou 350117, China
[4] The University of Tokyo, Meguro-ku, Tokyo 153-8505, Japan
yhren@fjnu.edu.cn

Abstract. In the Big Data era, we are surrounded by digital products and digital technology. The necessity for large-capacity storage equipment has increased. Holography data storage has been good candidate for a green data storage system, because of there are not only large storage capacities and high transfer rates, but also long life-time and low cost with the unique configuration. In this paper, we introduced the principle of the Holographic data storage system and its media structure. Some results of experimental and theoretical studies suggest that it is a very effective method. We also discussed some methods to increase the recording density and data transfer rates using phase and polarization modulations.

Keywords: Big Data, Holography, Polarization, Data Storage, Multi-Modulation.

1 Introduction

In the Big Data era there is an even higher demand for data storage technology. However, the main data storage technology is the magnetic storage technology. This kind of technologies not only has a limited life time (hard disk 5 years, tape 10 years), but also the data is stored in the two-dimensional plane of the media surface, so the storage density has no space for development. The solid-state storage technology represented by flash memory also faces the problems of limited storage density and charge loss. Holographic data storage is known as a high density, fast rate, long life and low cost method [1], but the actual storage density is far from its theoretical value. This paper intends to use the multi-dimensional modulated holographic data storage technology to explore a new mechanism to suppress holographic noise, study the inherent law of high-density preservation and holographic material structure, and solve the density problem of holographic data storage from a theoretical point of view.

S. Kurimura and T. Yagi (eds.), *Proceedings of Laser Display and Lighting Conference 2025*, Springer Proceedings in Physics 429,
https://doi.org/10.1007/978-981-96-9782-3_21

It's not difficult to modulate a holographic data page with amplitude, phase, and polarization, the question is how to demodulate them from the reconstructed data page. In the following, we will introduce the demodulation method of multi-dimensionally modulated information from two parts: amplitude and phase, i.e., the complex amplitude, and the demodulation of polarization and complex amplitude.

2 Complex amplitude modulation

In order to analyze the relationship between the diffraction pattern and the amplitude-phase data numerically, a complex amplitude demodulation method based on deep learning is proposed that can retrieve both the amplitude and phase from a single-shot near field diffraction intensity image [2]. By analyzing the correlation between the near field diffraction and the encoded data pages, the inverse problem of solving the complex amplitude from the intensity map is decomposed into two inverse problems of retrieving the amplitude and phase and represented by convolutional neural networks (CNN) respectively. After the CNNs' training is completed, the amplitude and phase data pages can be directly reconstructed from a diffraction intensity image.

3 Polarization multiplexing

Polarization holography can record the amplitude, phase and polarization information of light wave with polarization-sensitive materials [3]. This characteristic endows polarization holography with the advantage of boosting storage capacity and density specifically within the polarization dimension. A novel concept and construction method of Optical Polarized Orthogonal Matrix (OPOM) are proposed [4]. OPOM, an orthogonal matrix incorporating polarization elements, holds the potential to substantially elevate the modulation order in the polarization dimension. Experimental analysis has validated that crosstalk among these multi-channel holograms is negligible. This study demonstrates the viability of OPOM application in polarization holography, paving the way for the exploration of higher-order polarization orthogonality.

References

1. H. Horimai, X. Tan, and J. Li: Collinear holography. Appl. Opt. 44(13), 2575–2579 (2005).
2. J. Hao, X. Lin*, Y. Lin, M. Chen, R. Chen, G. Situ, H. Horimai and X. Tan*: Lensless complex amplitude demodulation based on deep learning in holographic data storage. Opto-Electronic Advances 6(3), 220157 (2023).
3. K. Kuroda, Y. Matsuhashi, R. Fujimura, and T. Shimura: Theory of polarization holography. Opt. Rev. 18(5), 374–382 (2011).
4. S. Zheng, J. Tan, X. Xu, H. Liu, Y. Yang, X. Lin, and X. Tan*: Optical polarized orthogonal matrix. Photonics Research 13(2), 373-381 (2025).

Dual-comb Spectroscopy Extended for Spatiotemporal Optical Sensing Applications using Structured Light with Orbital Angular Momentum

Akifumi Asahara[1] and Kaoru Minoshima[1]

[1] Graduate School of Informatics and Engineering, The University of Electro-Communications, 1-5-1 Chofugaoka, Chofu, Tokyo 182-8585, Japan
aki.asahara@uec.ac.jp, k.minoshima@uec.ac.jp

Abstract. Dual-comb spectroscopy (DCS) is one of the most attractive applications of optical frequency combs, because it realizes ultra-precise, broadband, and high-speed Fourier spectroscopy. The application of DCS has widely expanded to many fields, such as gas spectroscopy, distance measurement, material characterization. Among such techniques, the ultra-precise and highly controllable optical phase of combs have been utilized. However, the "phase" in this context is generally limited to the dimension of the longitudinal mode of light. Here, we focus on optical vortices with a spiral wavefront transporting orbital angular momentum (OAM). The spatial phase structure is applied to the dimensional conversion of optical combs into transverse modes. Recently, we proposed a fusion of these two concepts as an "optical vortex comb". This idea enables arbitrary spatiotemporal phase control based on the high controllability of the optical combs. In this presentation, we will demonstrate our recent progress on novel DCS techniques utilizing the optical vortex comb concept, for example, in-plane angle measurement and spatiotemporal spectroscopic applications. Through the proposed optical vortex comb concept, the applicability of the optical combs is greatly expanded far beyond the conventional area.

Keywords: Dual-comb spectroscopy, Optical frequency combs, Optical vortex

1 Concept of Dual-comb spectroscopy with optical vortex

Dual-comb spectroscopy (DCS) is one of the most representative applications of optical frequency combs, realizing ultra-precise, broadband, and high-speed Fourier spectroscopy [1]. The application of DCS has expanded widely in recent years. It has been applied to various targets, including gas spectroscopy, distance measurements, multidimensional spectroscopy [2], photon-level DCS [3], and other disciplines. Among such technical proposals, ultra-precise phase stability and high controllability of optical frequency combs have been utilized. However, the "phase" utilized here is generally limited to the dimension of the longitudinal mode of light, that is, the dimension of the direction of light propagation. If the phase controllability of optical combs can be extended to other dimensions, the applicability of DCS can be further extended.

S. Kurimura and T. Yagi (eds.), *Proceedings of Laser Display and Lighting Conference 2025*, Springer Proceedings in Physics 429,
https://doi.org/10.1007/978-981-96-9782-3_22

Here, we focused on optical vortices [4], which is a light with a spiral wavefront transporting orbital angular momentum (OAM). It is attracting attention as a new light source for technologies such as spatial multiplexing communication and spiral optical processing. The optical vortex has a characteristic phase distribution that changes along the direction of the azimuthal angle. The helicity of the optical vortex is characterized by a parameter of the Laguerre Gaussian beam, the topological charge number, ℓ. The spatial phase distribution of the optical vortex is expressed as $\ell\theta$, where θ is the azimuthal angle.

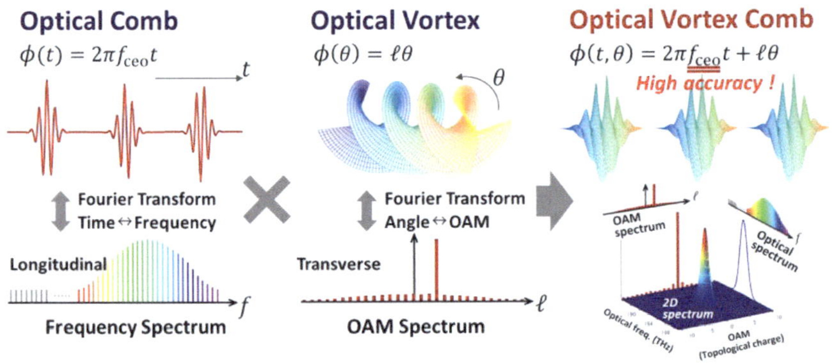

Fig. 1. Concept of the combination of optical frequency comb with optical vortex as optical vortex comb light.

This feature can be applied to the dimensional conversion of optical combs. Recently, we proposed a fusion of optical combs and optical vortices as an "optical vortex comb" concept [5,6], that was defined as the optical comb source that has OAM mode of optical vortex (Fig. 1). Because they have the freedom of orthogonal phase controllability in longitudinal and transverse dimensions, they have a great potential generating powerful benefits by combination of them. This idea enables arbitrary spatiotemporal phase control based on the high controllability of the optical comb. Through the optical vortex comb concept, the applicability of the optical combs is greatly expanded far beyond the conventional area.

In this presentation, we will present our recent progress on novel DCS techniques utilizing the optical vortex comb concept. For example, a measurement concept, "dual-vortex-comb spectroscopy [7], where it extends the precise and rapid phase measurement of DCS to the azimuth angle direction. The non-ambiguity angle measurement range is also investigated by changing the topological charges. The angle detected in our research is the azimuth angle in the beam plane corresponding to the "roll" axis. The angle information in the roll axis will allow us to access to unique measurement targets in the transverse plane, such as spinning objects and flow vorticity. Our developed methods will be useful in broad area, such as robotics, surface metrology, and motion tracking, LiDAR, and other advanced spectroscopic applications.

Through the proposed concept, the applicability of the optical combs is greatly expanded far beyond the conventional area.

This research was supported by MEXT/JSPS KAKENHI Grant Number 21H01840 and JST, PRESTO JPMJPR24L1.

References

1. I. Coddington, N. Newbury, and W. Swann, "Dual-comb spectroscopy," Optica **3**, 414–426 (2016).
2. B. Lomsadze and S. T. Cundiff, "Frequency combs enable rapid and high-resolution multidimensional coherent spectroscopy," Science **357**, 1389–1391 (2017).
3. N. Picqué and T. W. Hänsch, "Photon-level broadband spectroscopy and interferometry with two frequency combs," Proc. Natl. Acad. Sci. **117**(43), 26688–26691 (2020).
4. M. J. Padgett, "Orbital angular momentum 25 years on [Invited]," Opt. Express **25**(10), 11265–11274 (2017).
5. A. Asahara, S. Shoji, K. Kondo, Y. Wang, and K. Minoshima, " Coherent spatiotemporal control by combining optical frequency combs and optical vortices," CLEO:2018, STu4P.3 (2018).
6. A. Asahara and K. Minoshima, "Optical combs and optical vortices combined for spatiotemporal manipulation of light and matter," arXiv:2005.04705 (2020).
7. A. Asahara, S. Akiyama, and K. Minoshima, "Dual-comb spectroscopy for in-plane angle measurement using OAM vortex light," Opt. Express **31**(7), 11695-11704 (2023).

Design of Meta-Surface Lens for THz-wave Object Imaging System in Security Application

Seiya Suzuki, Rizadi Sasmita Darwis and Hiroshi Murata

Graduate School of Engineering, Mie University Tsu-City, Mie 514-8507 Japan.
421017@m.mie-u.ac.jp, murata@elec.mie-u.ac.jp

Abstract A new flat lens utilizing meta-surface structures was analyzed and designed for imaging applications operating in the THz frequency band. THz signal focusing characteristics with a relatively low-transmission loss were verified by use of three-dimensional field analysis. This lens is suitable for the application to three-dimensional under-cloth imaging systems for non-contact inspection for remote security applications in a public space.

Keywords: Meta-Surface, Lens, THz-wave, 3-D Imaging.

1 Introduction

Two- and/or three-dimensional imaging technologies utilizing THz-frequency band electromagnetic waves are promising for the applications in remote sensing and non-contact inspection for screening and checking in a public space [1], [2]. Advantages of THz-frequency wave imaging scheme are able to inspect an item or object covered by a cloth, leather, plastics, and paper indirectly and nondestructively, which are important to detect and to identify a dangerous item hidden in a pocket or under jacket. However, conventional radar-based sensing systems require a relatively long operational time to detect a hazardous object. Besides, rejections of unwanted background THz signals (thermal signals) are not so straightforward when using a standard thermal-based sensor. Therefore, it seems rather difficult to construct an imaging system with high throughput for the applications of screening and checking in crowded public places (such as a large sport stadium, arena, concert hall, train station, shopping mall, etc.) owing to the requirement with a high throughput for security checking operations.

Recently, we have proposed and developed a new THz/millimeter-wave (MMW) imaging technique based on the complex correlation and integral technique between the reflected signals from a target/object and the reference signal. By utilizing this imaging technique, we have demonstrated high-throughput indirect imaging systems to capture images with a high speed (>10 flames/s) [3]. In addition, by utilizing the optical lens-based technique to MMW imaging, the improvement of an effective antenna gain value for MMW signals and the magnification of the MMW image are obtainable [4]. A dielectric lens is also suitable to magnify THz images with this approach, however, a conventional lens for THz imaging is relatively bulky and heavy and not so straightforward to applications in practical system.

S. Kurimura and T. Yagi (eds.), *Proceedings of Laser Display and Lighting Conference 2025*, Springer Proceedings in Physics 429,
https://doi.org/10.1007/978-981-96-9782-3_23

In this paper, we report on an analysis and design of a new lens for THz imaging based on the meta-surface structure. By utilizing the tailored meta-surface structure, a compact light-weight lens for THz imaging can be obtained.

2 Basic structure of meta-surface lens

Fig. 1 shows the basic structure of the meta-surface lens and the unit cell structure of the meta-surface element [5]. A substrate is a transparent glass or plastic plate with a thickness of *T*. In the unit cell structure of the meta-surface, a pair of square metal patch plates with a length of *W*, and a pair of square metal rings with an outer length of *P*, and an inner length of *L* are fabricated on both the top and reverse-side surfaces of the substrate. These square metal patches and square metal rings are composed of a thin (~1 μm) metal film with good conductivity like Au, Cu, or Al, to cause an *LC*-circuit like resonance at the designed frequency. As a result, meta-surface operations are obtainable with a rather high transmission coefficient at the designed frequency range although the relatively large area of the substrate is covered with the metal elements.

The resonance frequency of the meta-surface element is strongly dependent on the size of the metal patch, *W*. Therefore, by changing only the metal patch size, *W*, gradually from the center to the outer of the lens, the phase front of the transmitted THz wave can be tuned, and the lens operation is obtainable at the deigned frequency.

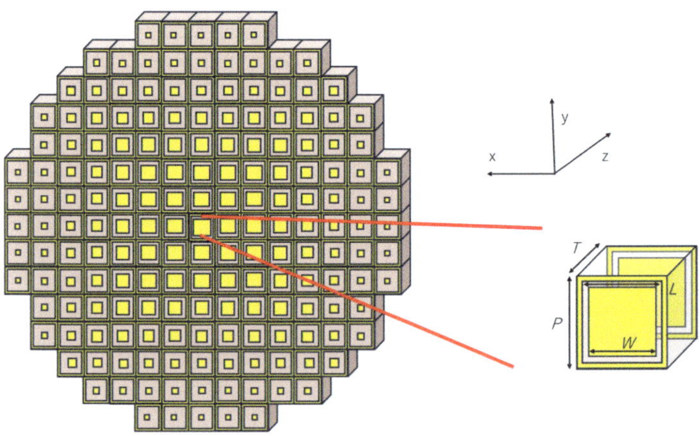

Fig. 1. Structure of meta-surface lens for THz-wave imaging.

3 Analysis of meta-surface

The transmission characteristics of the meta-surface structure were analyzed in detail by use of 3-dimensional field analysis software. The calculated transmission characteristics through the meta-surface structures with several W values (W = 600, 640, 680, and 720 μm) are shown in Fig. 2, where the substrate is assumed to be made of SiO_2 glass ($\varepsilon_r = 4.0$) with a thickness of T = 400 μm, and the metal film is assumed to be made of a thin Al (0.2 μm thickness), and the other meta-surface structure parameters are P = 1000 μm, and L = 900 μm. We can see that rather high transmissions are obtainable around the intrinsic resonance frequency of 80~100 GHz for these meta-surface structures as shown in Fig. 2 (a). Fig. 2 (b) indicates the calculated wavefront delay values of the transmitted THz waves just after the meta-surface structure.

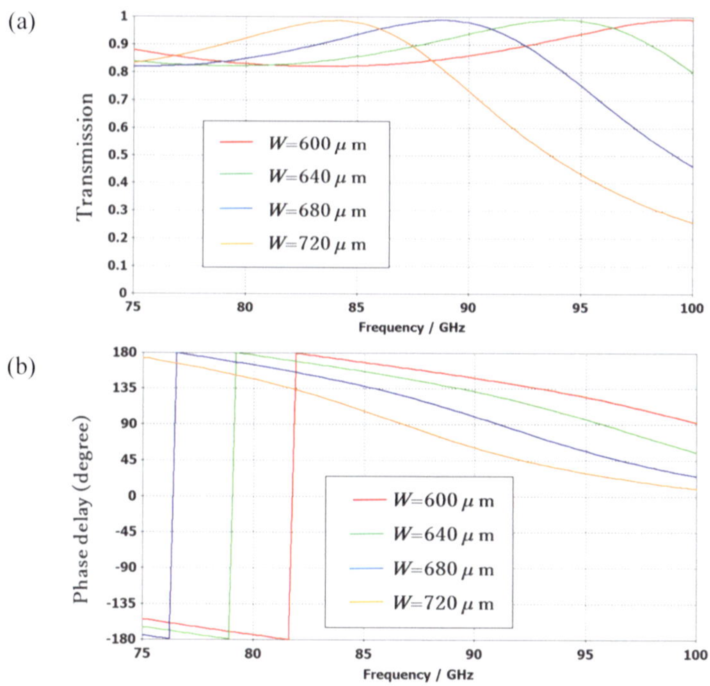

Fig. 2. Calculated transmission characteristics of meta-surface structure. (a) Magnitude and (b) phase of transmission coefficient.

4 Design for lens operation

By utilizing the phase delay characteristics of the THz waves passing through the meta-surface structure, we designed the THz lens. As a first trial, a cylindrical lens was designed and analyzed. The designed THz lens model and corresponding phase delay response are shown in Fig. 3, where 29 meta-surface elements were aligned vertically with changing the square metal length of W from 618 to 720 μm, where the phase delay values were tuned from 0 (at the center) to 51 degrees (both ends) gradually according to the square metal length of Ws. The lens height was 29 mm, and the operational frequency was set as 84 GHz.

(a)

Square metal length W	
W_0	720 μm
W_1	718 μm
W_2	716 μm
W_3	714 μm
W_4	712 μm
W_5	706 μm
W_6	700 μm
W_7	694 μm
W_8	688 μm
W_9	678 μm
W_{10}	658 μm
W_{11}	648 μm
W_{12}	638 μm
W_{13}	628 μm
W_{14}	618 μm

(b)

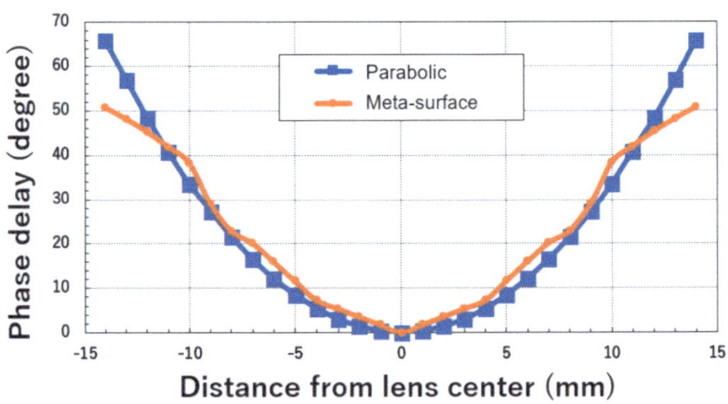

Fig. 3. (a) Designed meta-surface cylindrical lens. (b) Calculated phase delay characteristics (operational frequency 0.84 THz).

The focusing characteristics of the designed meta-surface cylindrical lens were analyzed by use of 3-dimentional field analysis software also. Fig. 4 shows the calculated field variations passing through the meta-surface lens. We can see that the THz field wave-front is converging and the focusing lens is about 120 mm. An estimated gain value is about 3.9 dB. By utilizing the large aperture size lens structure (Diameter size of ~300 mm), the lens antenna gain of > 30 dBi is expected.

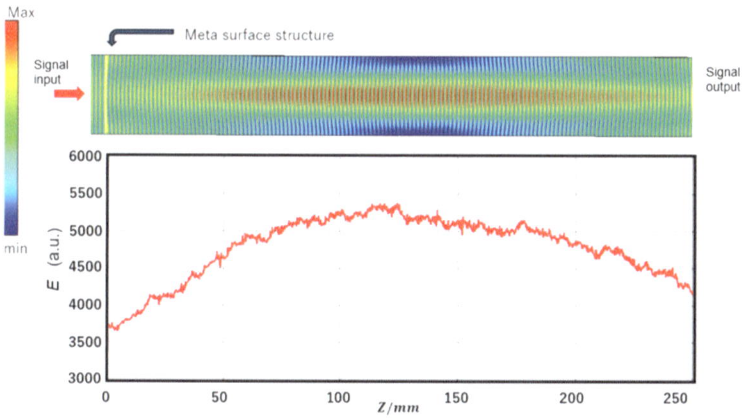

Fig. 4. Calculated focusing characteristics of the designed meta-surface lens (operational frequency 0.84 THz).

5 Conclusions

The analysis and design of the new meta-surface lens are reported. A flat lens with antenna gain over 30 dBi is expected by use of the simple meta-surface structure. Another important point is light weight compared to conventional lenses of PTFE or other dielectric materials. We are now trying to fabricate the designed lens to check the focusing and imaging characteristics.

Acknowledgements

The authors thank Dr. Yui Otagaki and Mr. S. Sato of Mie University, Japan, for their support in the calculation. This work was supported in part by the research program from the Ministry of Land, Infrastructure, Transport and Tourism (MLIT), Japan.

References

1. Z. Wang, T. Chang, and H.-L. Cui, "Review of Active Millimeter Wave Imaging Techniques for Personnel Security Screening," IEEE Access, Vol. 7, 148336-148350 (2019).
2. S. S. Ahmed, A. Schiessl, F. Gumbmann, M. Tiebout, S. Methfessel and L.-P. Schmidt, "Advanced Microwave Imaging," IEEE Microwave Magazine, Vol. 13, No. 6 (2012).
3. N. Yonemoto, A. Kohmura, S. Futatsumori, K. Morioka, and Y. Makita, "Passive Radio Imaging of Hybrid Radar System for Security Inspections," 17th European Radar Conference (EURAD 2020), pp. 378-381, Utrecht, Netherlands, Jan. 2021.
4. A. Setiawan, A. Yamawaki, N. Yonemoto, H. Nohmi, and H. Murata, "Millimeter-Wave Imaging Using Dielectric Lens for Security Application," 19th European Radar Conference (EuRAD 2022), EuRAD14-3, pp. 225-228, Milan, Italy, Sep. 2022.
5. Q. Yang, J. Gu, D. Wang, X. Zhang, Z. Tian, C. Ouyang, R. Singh, J. Han, and W. Zhang, "Efficient flat metasurface lens for terahertz imaging," Optics Express, Vol. 22, No. 21, 25931, 2014.

2- and 3-Dimensional Imaging Using Field Interference Integration for Remote Sensing and Security Inspection Applications

Yamato Iwata, Riyogo Ishida, Arie Setiawan, Shu Sato, and Hiroshi Murata

Mie University 1577 Kurima-Machiya-Cho, Tsu-City, Mie 514-8507 Japan.
murata@elec.mie-u.ac.jp

Abstract. Millimeter-wave imaging is attractive as an indirect inspection technique for security solutions due to its ability to penetrate common cloth, leather, plastic, and paper to detect concealed objects without harm. This paper reports experimental results of 2- and 3-dimensional imaging results utilizing the millimeter-wave interference integration technique and a dielectric lens. Spatial filtering techniques of measured images are also discussed.

Keywords: Millimeter-wave imaging, Security inspection, Image projection, Spatial filter.

1 Introduction

Radar and imaging technologies utilizing millimeter-wave (MMW) or THz-frequency band electromagnetic waves are important for the remote sensing and noncontact screening systems for public space [1]-[3]. MMW/THz-frequency waves can penetrate a cloth, leather, plastics, and paper to identify a hidden hazardous object covered with these materials. However, conventional radar-based sensing systems usually require a relatively long operational time to generate an image and to detect a hazardous object. Therefore, it seems rather difficult to construct and to install an imaging system for security checking in public places (such as a stadium, arena, concert hall, train station, shopping mall, etc.) owing to the requirement of a high throughput operation.

Recently, we have proposed and developed a new MMW imaging technique based on the complex correlation between the reflected signals from a target/object and the reference signal, which enables to have a high-throughput operation compared to the conventional MMW imaging systems [4]-[6]. In addition, by utilizing the optical lens-based technique to MMW imaging, the improvement of antenna characteristics for MMW signals and the magnification of the MMW image are obtainable [5], [6]. A fusion with laser-based displays is also applicable to display generated images.

In this paper, experimental results for generating 2-/3-dimensional images utilizing MMW signals are reported. By utilizing a frequency sweep operation with the complex correlation scheme, 2-/3-dimentional images of a test target are successfully generated.

S. Kurimura and T. Yagi (eds.), *Proceedings of Laser Display and Lighting Conference 2025*, Springer Proceedings in Physics 429,
https://doi.org/10.1007/978-981-96-9782-3_23

2 Imaging System

Figure 1 shows the basic configuration of the imaging system with the field complex correlation technique [5], [6]. A MMW signal in the frequency range of 75~110 GHz (W-band) was generated from a signal source in a vector network analyzer (VNA). This signal was supplied to a TX antenna (standard gain horn antenna with a gain of ~20 dBi) and irradiated to a target to be measured. The emitted signal was reflected by a target and the reflected signal was transferred to a RX antenna through a dielectric lens. This RX antenna was scanned in the measurement plane along the *x*- and *y*-axes by use of a mechanical scanner to obtain 2-dimentional information. Then, the received signals were converted to the lower (IF) frequency range and analyzed by use of the VNA to obtain the field complex correlation between the measured signal and the reference signal.

Two dielectric lenses with a large aperture size (a diameter of 250 mm or 500 mm) were used in the experiment for concentrating the signal reflected by a target. The lens was composed of PTFE. A typical measurement length (d_1 and d_2) is 500~1000 mm.

To obtain 3-dimentional images from the measurement data, the signal frequency was scanned from 75 to 110 GHz with a phase-locked condition (i.e. with a completely coherent condition), and the received signals in the frequency domain were converted to the time domain responses by use of the Fourier transformation. As a result, the information along the Z-direction can be obtained as the time delay in each measurement position in the X-Y plane, and 3-dimentional images can be obtained.

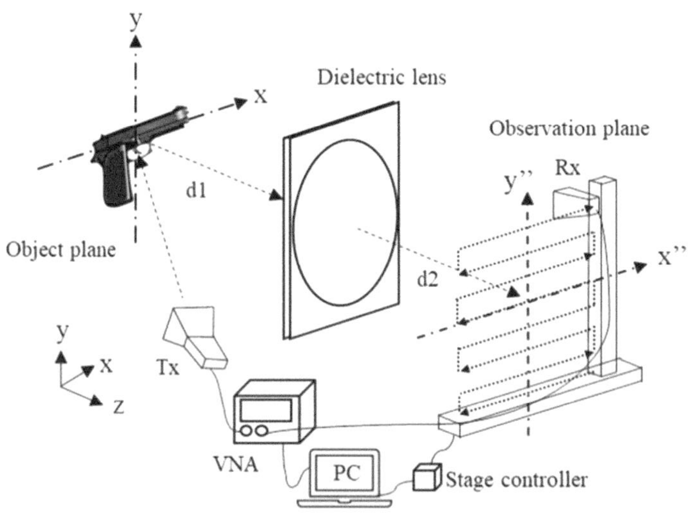

Fig. 1. MMW imaging system utilizing dielectric lens.

3 Results and Discussions

An example of the generated 3-dimentional image for a test object of a mannequin with a plastics gun is shown in Fig. 2. Details of the test object are also shown in Fig. 3. We can see that a clear 3-dimentional image of the plastic gun is generated successfully through a covered cloth (T-shirt). Fig. 4 shows the 2-dimentional (sliced) images obtained by slicing the 3-dimentional image of Fig. 2. We can see that the shape of the hidden target was identified successfully. Detailed analysis repeating measurements are now underway.

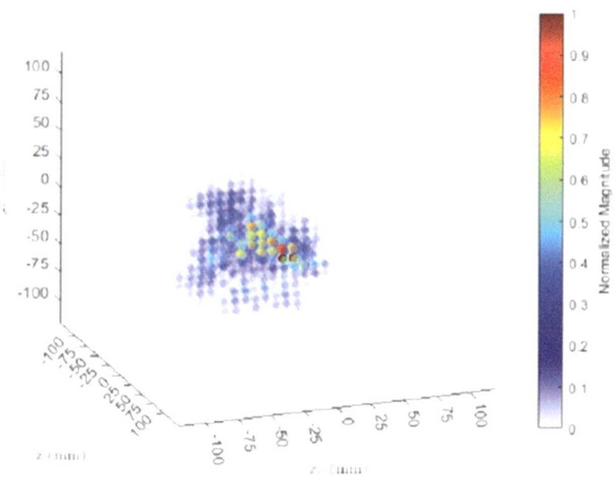

Fig. 2. Generated 3-dimentional MMW image of a test object covered with T-shirt.

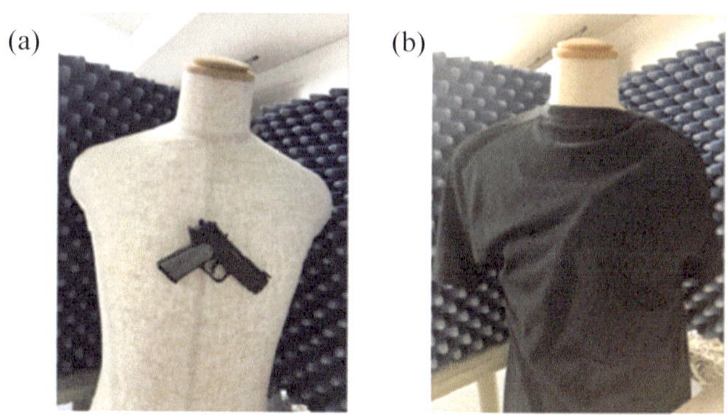

Fig. 3. Details of a test object of a mannequin and a plastics gun. (a) Without T-shirt cover, (b) With T-shirt.

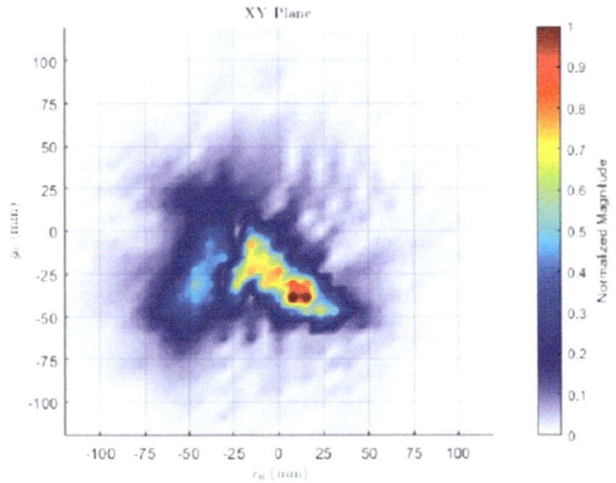

Fig. 4. Generated 2-dimentional MMW image.

Our system is essentially a kind of Fourier optics-based imaging system. Therefore, several imaging processing techniques are appliable, for example, spatial frequency filtering, correlation, pattern matching, and so on. For image magnifications, a meta-surface structure-based lens or filter is also applicable.

Another interesting approach is a fusion with laser display technologies for generating and projecting 3-/2-dimensional images measured by the MMW technique. Details will be discussed at the conference.

Acknowledgements

The authors thank Dr. Yui Otagaki of Mie University, Japan, Dr. Hitoshi Nohmi of Alouette Technology Inc, Japan, and Dr. Naruto Yonemoto of National Institute of Maritime, Port and Aviation Technology, Japan for their help and support in the experiments. This work was supported in part by the research program from the Ministry of Land, Infrastructure, Transport and Tourism (MLIT), Japan.

References

1. Z. Wang, T. Chang, and H.-L. Cui, "Review of Active Millimeter Wave Imaging Techniques for Personnel Security Screening," IEEE Access, Vol. 7, 148336-148350 (2019).
2. S. S. Ahmed, A. Schiessl, F. Gumbmann, M. Tiebout, S. Methfessel and L.-P. Schmidt, "Advanced Microwave Imaging," IEEE Microwave Magazine, Vol. 13, No. 6 (2012).
3. K. Mizuno, Y. Wagatsuma, H. Warashina, K. Sawaya, H. Sato, S. Miyanaga, Y. Yamanaka, "Millimeter-Wave Imaging Technologies and Their Applications," 2007 IEEE International Vacuum Electronics Conference, 15-17 May 2007.

4. N. Yonemoto, A. Kohmura, S. Futatsumori, K. Morioka, and Y. Makita, "Passive Radio Imaging of Hybrid Radar System for Security Inspections," 17th European Radar Conference (EURAD 2020), pp. 378-381, Utrecht, Netherlands, Jan. 2021.
5. A. Setiawan, A. Yamawaki, N. Yonemoto, H. Nohmi, and H. Murata, "Millimeter-Wave Imaging Using Dielectric Lens for Security Application," 19th European Radar Conference (EuRAD 2022), EuRAD14-3, pp. 225-228, Milan, Italy, Sep. 2022.
6. A. Setiawan, N. Yonemoto, H. Nohmi, and H. Murata, "Experimental Investigation of Millimeter-Wave 3D Image Projection Using Dielectric Lens for Security Application," 20th European Radar Conference (EuRAD 2023), Berlin, Germany, Sep. 2023.

Visible light generation by sum frequency conversion for quantum sensing and imaging

Sunao KURIMURA[1,2], Yuta KOCHI[2], Junko ISHI-HAYASE[2]

[1] National Institute for Materials Science, Tsukuba 305-0044, JAPAN
[2] Keio University, Yokohama 223-8522, JAPAN
skurimura@nifty.com

Abstract. We demonstrate high-speed optical sampling of time-bin qubits through visible-light generation via sum-frequency generation (SFG) using femtosecond laser pulses. The experiment utilizes a periodically poled Mg-doped stoichiometric lithium tantalate (PPMg:SLT) crystal to convert near-infrared signal pulses (1520 nm) and pump pulses (820 nm) into visible light (533 nm). We achieved a conversion efficiency of 10.1% and a temporal resolution of 415 fs, while preserving quantum information, with potential applications in time-gated quantum sensing, LiDAR, and high-precision imaging.

Keywords: Optical sampling, sum-frequency generation, quasi-phase matching

1 Introduction

Laser display technologies incorporate diverse components such as light sources, scanners, modulators, sensors, and image processors, and they have significantly advanced both scientific and industrial applications. Building on this technological foundation, the Laser Display Conference (LDC) has progressively expanded its scope to include applications in sensing and imaging, leveraging the shared core technologies that underpin these domains.

Laser sensing technologies play a crucial role not only in display applications, but also in imaging and high-precision measurement. In particular, nonlinear optical wavelength conversion has gained attention as an advanced method for high-speed sensing through visible light generation. The nonlinear parametric process uses virtual transitions of excited electrons, exhibiting ultrafast response [1].

Laser sensing with quantum-entangled photons is highly promising for high-precision measurements, as demonstrated in quantum optical coherence tomography (Q-OCT). The entanglement brings great advantage of dispersion-free broadband OCT, leading to high-depth resolution in tomography [2].

This study focuses on high-speed optical sampling of qubit, through visible light generation via sum frequency generation (SFG) using femtosecond laser pulses. By utilizing a periodically poled Mg-doped stoichiometric lithium tantalate (PPMg:SLT) crystal [3], we successfully achieved up-conversion-based optical sam-

S. Kurimura and T. Yagi (eds.), *Proceedings of Laser Display and Lighting Conference 2025*, Springer Proceedings in Physics 429,
https://doi.org/10.1007/978-981-96-9782-3_25

pling from near-infrared signal light (1520 nm) and femtosecond pump light (820 nm) to visible light at 533 nm.

2 Sum-Frequency Generation for high-speed sampling

The experimental setup consists of a femtosecond Ti:Sapphire laser (820 nm, 200 fs, 76.3 MHz) as the pump source and a near-infrared optical parametric oscillator (1520 nm, 240 fs) as the signal source. These pulses are combined in a PPMg:SLT bulk crystal, where sum-frequency generation occurs, producing visible light at 533 nm for a mutual correlation function. The schematic view is illustrated in Figure 1 [4].

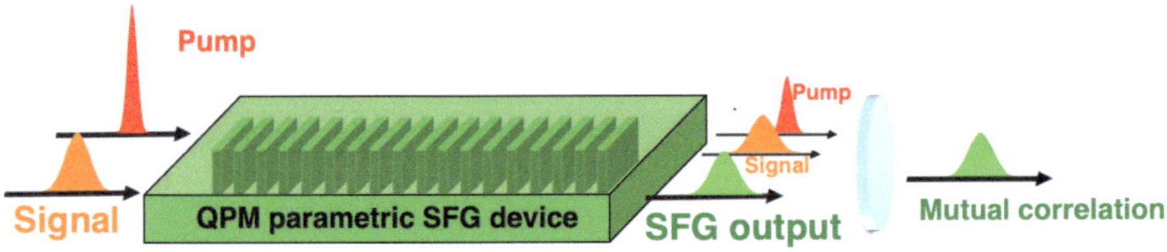

Fig.1 High-speed optical sampling through visible light generation

To analyze the group-velocity dispersion in SFG, we tested PPMg:SLT crystals with three different lengths: 1 mm, 2 mm, and 3 mm. Through careful optimization of the crystal length and pump conditions, we achieved a conversion efficiency of 10.1% and a temporal resolution of 415 fs with a 2-mm-long nonlinear crystal. These results demonstrate the potential of SFG-based visible light generation for laser sensing while keeping quantum information of the qubit, offering advantages in both high-resolution time-gating and ultrafast optical detection. Phase retrieval from time-bin qubit pulses will be demonstrated in the presentation.

The visible light generated by this SFG system provides high temporal resolution, making it well-suited for ultrafast laser sensing, high-precision imaging, LiDAR, and remote sensing.

References

1. L.K. Oxenløwe, F. G. Agis, C. Ware, S. Kurimura et al., "640-Gbit/s Data Transmission and Clock Recovery Using an Ultrafast Periodically Poled Lithium Niobate Device," J. of Lightwave Tech., 27, pp.205-213 (2009).
2. M. Okano, H. H. Lim, R. Okamoto, N. Nishizawa, S. Kurimura, S. Takeuchi, "0.54 μm resolution two-photon interference with dispersion cancellation for quantum optical coherence tomography," Sci. Rep. 5, 18042 (2016).
3. O. Kuzucu, F.N.C. Wong, S. Kurimura, S. Tovstonog, "Time-resolved single-photon detection by femtosecond upconversion," Opt. Lett., 33, pp.2257-2259 (2008).
4. Y. Kochi, S. Kurimura, and J. Ishi-Hayase, "Optimization of crystal length in a pulse-pumped up-conversion single-photon detector for decoding femtosecond time-bin qubits," Opt. Express 32, pp. 47549-47560 (2024).

Volumetric Displays with Laser-Excitation: From Fingertip to Fist-Sized Graphics and Color

Kota Kumagai [0009-0005-6573-451X] and Yoshio Hayasaki [0000-0001-6826-9995]

Utsunomiya University, 7-1-2 Yoto, Utsunomiya, Tochigi 321-8585 Japan
kumagai@a.utsunomiya-u.ac.jp

Abstract. We introduce recent advances in volumetric displays based on femtosecond laser excitation in terms of graphics scale and color. Centimeter-scale graphics with sufficient brightness for human observation have been achieved. Furthermore, a rendering method involving the generation of light scattering points and controlling the color of illumination, which is promising for color graphics, has been demonstrated.

Keywords: Volumetric display, Femtosecond laser, Computer-generated hologram.

1 Introduction

Volumetric displays render 3D volumetric graphics by generating emission points and light scattering points as voxels [1]. A volumetric graphics gives users a natural depth perception as if they were looking at an object in physical space because the points are arranged to form the shape of the target object. However, many challenges still remain in the research field of volumetric displays, such as the method for generating emission and light scattering points, the number of points, the scale of graphics, and color.

Toward the realization of volumetric displays, we have been developing a rendering method based on femtosecond laser excitation, beam scanning, and focal point design using computer-generated holograms (CGHs). Our group has investigated the relationship between the screen material and the excited point, and demonstrated its application to graphics rendering. Emission points obtained by air excitation have enabled augmented reality representations, in which aerial graphics are directly superimposed onto objects [2]. Microbubbles with slow rising velocity obtained by excitation of glycerin enabled the rendering of volumetric bubble graphics by stacking 2D cross-sections [3]. Emission points obtained using different screen structures, such as stacked polymer plates with different laser dye in each layer [4] and structured voxels consisting of three sub-voxels with RGB emission colors [5], were combined with multiple focal points by CGH to demonstrate volumetric graphics rendering and its colorization.

In this presentation, we introduce recent advances in volumetric display technologies for achieving larger scale graphics and color representation. Centimeter-scale graphics were successfully rendered by reconstructing an optical system that achieves both large scale and sufficient brightness for human observation [6, 7]. Furthermore, a

S. Kurimura and T. Yagi (eds.), *Proceedings of Laser Display and Lighting Conference 2025*, Springer Proceedings in Physics 429,
https://doi.org/10.1007/978-981-96-9782-3_26

method for generating light scattering points using laser-driven condensation has been developed, which is promising for color rendering.

2 Results

Figure 1 (a) shows a sphere-shaped volumetric graphics with 1500 voxels rendered by stacking 14 circles. The graphics was captured with an exposure time of 30 s. We successfully rendered a volumetric graphics with an axial length of 4.2 cm. Figure 1 (b) shows a time series of images rendered using droplets formed through laser-driven condensation in ethanol supersaturated vapor. We have demonstrated color graphics by dynamically switching the wavelength of the coaxially incident RGB laser illumination.

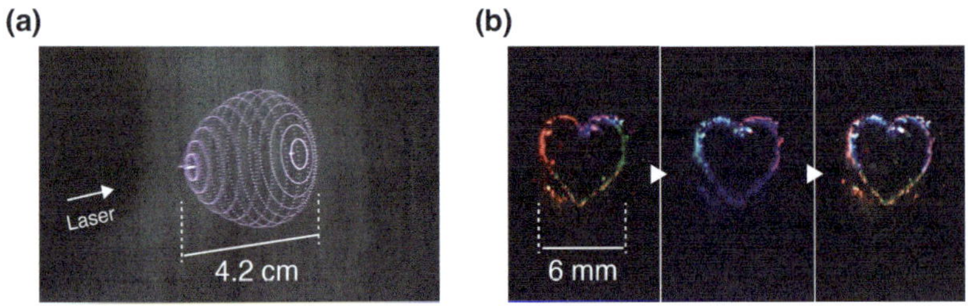

Fig. 1. (a) Aerial volumetric graphics with centimeter-scale. (b) Time series of heart-shaped graphics with color.

Acknowledgement

This work was supported by JSPS KAKENHI Grant Number JP 25K03148. We would like to thank Keisuke Numazawa, Tatsuki Mori, Hisashi Oka, Kazuki Horikiri and Tetsuji Suzuki for their contribution to the acquisition of experimental data.

References

1. B. G. Blundell et al., "Volumetric Three-Dimensional Display Systems," (Wiley-IEEE Press, USA, 2000).
2. Y. Ochiai et al., "Fairy Lights in Femtoseconds: Aerial and Volumetric Graphics Rendered by Focused Femtosecond Laser Combined with Computational Holographic Fields," ACM Transactions on Graphics **35**, 17:1–17:14 (2016).
3. K. Kumagai et al., "Volumetric bubble display," Optica **4**, 298-302 (2017).
4. K. Kumagai et al., "Volumetric display with holographic parallel optical access and multilayer fluorescent screen," Optics Letters **40**, 3356–3359 (2015).
5. K. Kumagai et al., "Three-dimensionally structured voxels for volumetric display," Optics Letters **43**, 3341-3344 (2018).
6. K. Kumagai et al., "Fist-sized aerial volumetric display with femtosecond laser drawing," J Soc Inf Display. https://doi.org/10.1002/jsid.2025 (2025).
7. K. Kumagai et al., "Volumetric Display with Dual-Path Holographic Laser Rendering," ACM SIGGRAPH 2024 Emerging Technologies, 1-2 (2024).

Holographic Optical Elements for Application in Wound Healing Monitoring

Pamela Stoeva,[1,2] Tatsiana Mikulchyk, [1,2] Dervil Cody, [1,2] Maria Antonietta Ferrara,[3] Giuseppe Coppola,[3] Suzanne Martin, [1,2] and Izabela Naydenova, [1,2]*

[1]Centre for Industrial and Engineering Optics, School of Physics, Clinical and Optometric Sciences, Technological University Dublin, D07 ADY7 Dublin, Ireland
[2]Physical to Life Sciences Research Hub, Technological University Dublin13 Camden Row, D08 CKP1 Dublin, Ireland
[3]Institute of Applied Sciences and Intelligent Systems (ISASI-NA), Via Pietro Castellino 111, 80131 Naples, Italy
* Izabela.naydenova@tudublin.ie

Abstract. Development of cost-effective, user friendly and readily accessible sensors for wound healing monitoring is needed to overcome growing global healthcare challenges associated with wound management. Such devices will extend basic healthcare to remote areas on earth and support long-term space missions by facilitating on-the-spot wound assessment. The fundamental biochemical processes associated with wound healing are typically related to parameters such as dissolved oxygen levels, pH, temperature, and lactate concentration. Real-time quantification of these parameters enables the assessment of local wound recovery and facilitates wound management. We present two sensor systems that utilise Holographic Optical Elements (HOEs) as promising options due to their diverse sensing capabilities, advantageous physical properties (e.g., lightweight and flexibility), and suitability for mass manufacturing.
We present two types of HOEs – a holographic coupler for future use in dissolved oxygen sensing system and a holographic grating for use in a temperature sensor in the temperature range of 33-37°C, relevant to wound healing.

Keywords: holographic optical elements, holographic sensors, wound healing monitoring, temperature sensor, holographic coupler.

1 Introduction

Chronic wounds represent a significant healthcare challenge, for example it is affecting approximately 3% of the U.S. Medicare population and incurring annual costs between \$28.1–\$97.4 billion [1]. Similarly, the UK's NHS allocates £8.3 billion per year to wound care, with untreated wounds increasing costs by 133% [2]. Poor wound management can lead to severe complications, including infections and amputations, with one occurring every 20 seconds globally, significantly reducing patients' quality of life [1]. Current wound assessment techniques often rely on costly, intricate procedures that hinder timely and accessible monitoring, particularly in resource-limited

S. Kurimura and T. Yagi (eds.), *Proceedings of Laser Display and Lighting Conference 2025*, Springer Proceedings in Physics 429,
https://doi.org/10.1007/978-981-96-9782-3_27

settings. Given the projected $23.2 billion wound care market by 2030, there is a growing demand for innovative, cost-effective, and scalable sensing solutions (Fig. 1) [1].

Holographic optical elements offer a compact (each individual element could be as small as a few mm²), reliable, and cost-effective alternative for direct optical feedback, reducing reliance on expensive instrumentation while enabling practical real-time assessment of wound conditions. The incorporation of holographic multi-waveguide systems, integrating optical couplers for dissolved oxygen sensing and temperature-sensitive photopolymer gratings, enables real-time monitoring of critical wound biomarkers [3]. This advancement will support continuous, non-invasive monitoring, improving early detection of complications, streamlining clinical decision-making, and enhancing patient outcomes through proactive, data-driven wound management in both clinical and remote settings.

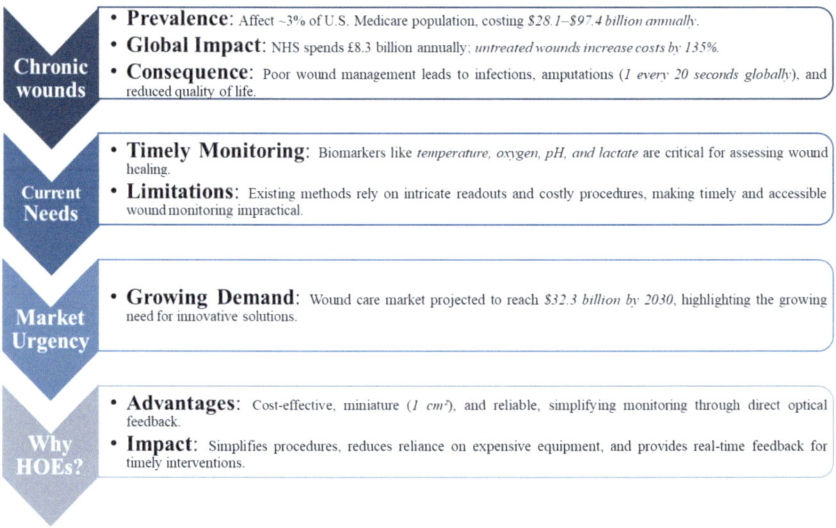

Fig. 1. Factors justifying the need for innovative wound healing monitoring [1-3]

2 Principle of operation

2.1 Holographic coupler

The holographic coupler enables precise light delivery and extraction through slanted volume phase transmission gratings, guiding an excitation beam into the sensing region and collecting luminescence-based signals for oxygen detection. Unlike conventional coupler fabrication requiring bulky optical elements, this system leverages a shorter recording wavelength to predefine the Bragg recording angle, ensuring efficient total internal reflection coupling at operational wavelengths Fig. 2 [4,5]. Multiplexed coupler gratings, fabricated via angular and peristrophic multiplexing, will allow parallel detection from multiple sensing sites, enhancing measurement flexibility. Optimising the Bragg condition for specific operational directions and wavelengths ensures selective and precise light coupling and guidance, supporting a compact, integrated sensing platform for real-time fluorescence-based wound monitoring.

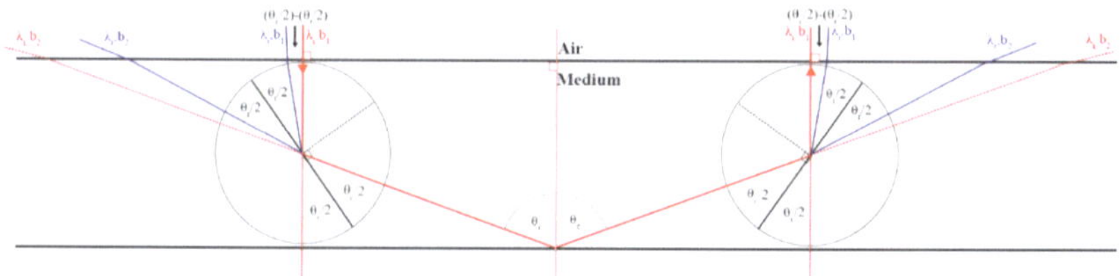

Fig. 2 Schematic presentation of the operational (red) and recording (blue) beams optical paths.

2.2 Temperature sensor

The temperature sensitive HOE functions as an optical transducer, where holographic photopolymer gratings undergo a Bragg angle shift due to thermally induced changes in refractive index and material dimensions as shown in Fig. 3 [6,7]. These structural modifications alter diffraction efficiency, leading to Bragg angle detuning, where deviation from the initial Bragg-matched condition results in a measurable decrease in diffracted beam intensity. This effect enables a direct optical readout of temperature variations within the clinically relevant 33–37°C range [8]. The photopolymer composition is tailored for reversible and repeatable diffraction responses, ensuring stable long-term monitoring. To minimise environmental interference (such as interaction with the wound exudate), the sensor is laminated, reducing external sensitivity while preserving temperature selectivity.

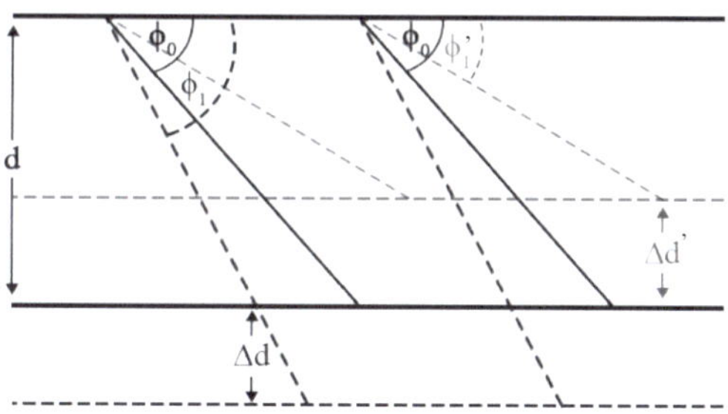

Fig. 3 Change of slant angle due to dimensional changes of the layer.

3 Materials and methods

3.1 Materials

Two types of materials were used in this study – an environmentally stable photosensitive solg-gel material and temperature sensitive photopolymer material. The sol-gel process is a chemical route for synthesising inorganic-organic and hybrid materials through the hydrolysis and condensation of metal alkoxides or organosilane precursors. Class II hybrid systems, such as the photopolymerisable hybrid sol-gel (PHSG) material used in this study, feature covalent bonds between organic and inorganic networks, enhancing structural integrity and chemical resistance, including water resistance of up to 1000 hours [9,10]. These properties make PHSG ideal for demanding optical applications where long-term durability is required (Table 1). Additionally, the material exhibits a highly efficient 45-minute curing time, significantly reducing fabrication complexity compared to conventional sol-gel and photopolymer approaches. The PHSG system consists of a silica-based matrix chemically linked to polymerisable organic groups, enabling direct holographic structuring of volume gratings with consistent and reliable optical efficiency.

Table 1. Key optical and mechanical characteristics of the sol-gel materials utilized in the fabrication of the holographic couplers [9. 10]

Spectral sensitivity	380-540 nm (CW) 780 nm (pulsed)
Resolving power	6000 lines/mm
Photosensitive layer thickness	20-300 um
Substrate	glass
Diffraction Efficiency	Up to 100 % (transmission)
Refractive index modulation	3×10^{-3} (CW) 9×10^{-3} (pulsed)
Water resistance	Up to 1000 hours
Temperature resistance	60 min at 130 °C
Other relevant properties	Biocompatible, stable, robust, fast layer preparation

In contrast to PHSG, which priorities stability and long-term robustness, photopolymers are organic materials that undergo photoinduced polymerisation, where exposure to spatially modulated light initiates cross-linking reactions, forming a spatially modulated polymer network [11]. This process induces refractive index modulation, enabling the recording of efficient volume phase holographic gratings while allowing for tunable optical responses, making them suitable for sensing applications. Unlike sol-gel-based systems, photopolymers have been extensively studied for their chemical tunability and capacity for selective interactions, enabling material formulations to be tailored for specific analytes or sensing conditions [12–14]. Their recording dynamics, governed by fast polymerisation kinetics and suitability to record controlled diffraction efficiency, make them well-suited for optical sensing applications, particularly where precise and long-term material selectivity is required. Through copolymerisation or adjustments in chemical composition, photopolymers can be engineered to function within a defined response range, ensuring high sensitivity and specificity for targeted sensing applications.

3.2 Holographic recording

The recording of PHSG slanted volume transmission gratings for operation as a holographic coupler is performed using a two-beam interference setup. A 476.5 nm Argon ion laser provides the recording beam with suitable wavelength, generating an interference pattern with a spatial frequency of 1720 ± 5 lines/mm, determined by a 56.4° ± 0.2° angle between the interfering beams. This configuration results in a 21.5° slant within the medium for a 633 nm coupler and a 23° slant for a 700 nm coupler. To ensure consistent high diffraction efficiency, the exposure intensity and duration are optimised to 3 mW/cm^2 and 20 s, respectively, yielding an average diffraction efficiency of 60 % after post recording stabilisation with a 532 nm beam (3 mW/cm^2 at 300 s).

For the fabrication of multiplexed couplers, sequential exposures are carried out at different and/or mirrored angles, employing angular and peristrophic multiplexing techniques. This approach enables the recording of multiple gratings within the same location of the layer, allowing for simultaneous signal guidance and detection across various sensing sites while maintaining the same Bragg angle.

Beyond their use in optical waveguiding, slanted volume transmission gratings can also be engineered for sensing applications, such as temperature monitoring. The recording of slanted volume transmission gratings in the temperature sensitive photopolymer was performed using a two-beam interference setup with a 532 nm laser. A beam splitter produced two equal-intensity beams that interfered in the photopolymer layer, forming an interference pattern with a spatial frequency of 820 ± 5 lines/mm, set by a 10° ± 0.05° angle in the medium. To achieve the required slanted grating geometry, the recording alignment was rotated by 15° ± 0.05° in air, adjusting the fringe orientation within the material. The total recording intensity was 5 mW/cm^2, with exposure duration controlled via an electronic shutter, leading to an initial diffraction efficiency of 50 ± 10%, ensuring operation within a linear Bragg detuning regime. After recording, the layers underwent UV curing for 20 minutes at 20 mW/cm^2 to enhance stability, followed by lamination with a 125 µm LEXAN (8010MC) film to minimise environmental sensitivity while preserving temperature responsiveness.

4 Results and discussion

4.1 Holographic coupler

The holographic coupler successfully facilitated both in-coupling and out-coupling of optical signals, demonstrating its effectiveness in guided wave-based sensing (Fig. 4a). In-coupling was validated by directing a 633 nm laser beam onto the coupler at the operational input angle, resulting in reconstructed propagation at the total internal reflection angle, ensuring efficient light confinement within the waveguide system. Out-coupling was confirmed by showing that a signal propagating at an angle equal to or greater than the angle for total internal reflection met the Bragg condition for an out-coupler along its axis, enabling controlled light extraction at the predefined output angle.

For fluorescence signal detection, a secondary HOE was implemented to operate at a different wavelength. A photoluminescent chamber, excited at 633 nm, emitted fluorescence at 700 nm (Fig. 4b), which was subsequently out-coupled via the secondary HOE. This selective out-coupling approach enabled efficient signal extraction. The fluorescence signal was successfully guided to an optical fibre spectrometer for detection, confirming the functional integration of the in- and out-coupling gratings. The ability of the coupler to facilitate both excitation and emission light pathways highlights its viability for sensing and spatial mapping through coupler multiplexing.

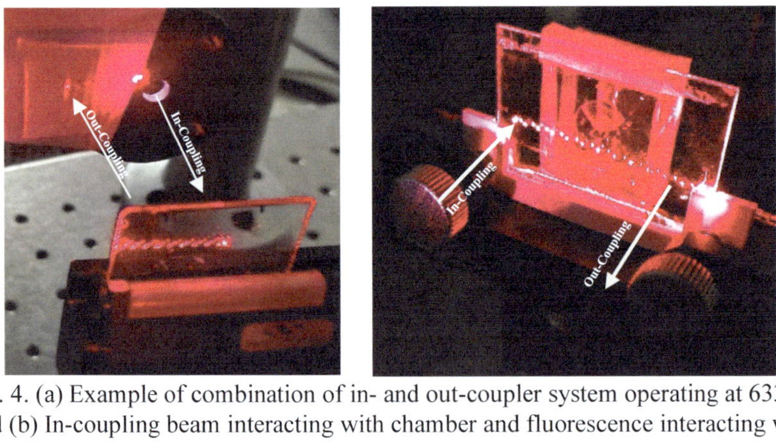

Fig. 4. (a) Example of combination of in- and out-coupler system operating at 633nm and (b) In-coupling beam interacting with chamber and fluorescence interacting with out-coupler.

4.2 Temperature sensor

The temperature response of the holographic grating sensor was characterised through controlled heating cycles within the clinically relevant range of 33–37 °C. The diffraction efficiency of the first-order diffracted beam was monitored in real-time to evaluate Bragg detuning, caused by thermally induced changes in refractive index and layer dimensions. To ensure accurate comparison across samples, the normalized diffracted beam intensity was used, mitigating variations in initial DE. Temperature cycling was performed at 0.5 °C/min, with each increment held for 120 s to assess stability before progressing to the next step (Fig 5). The material demonstrated a repeatable response across multiple cycles, confirming the reversibility of the diffraction efficiency shift.

Among the tested compositions, a specific formulation (D1: 25% acrylamide (AA): 75% diacetone acrylamide (DA): 25% N,N′-methylenebis(acrylamide) (BA)) exhibited the highest sensitivity, with a normalised intensity change of 18%/°C, representing a fivefold improvement over previously studied compositions. Additionally, the material maintained a stable response over 10 consecutive cycles in 20-day-old samples, demonstrating long-term stability for wound temperature monitoring. These results confirm the feasibility of holographic photopolymer gratings as optical transducers for non-invasive, continuous temperature sensing.

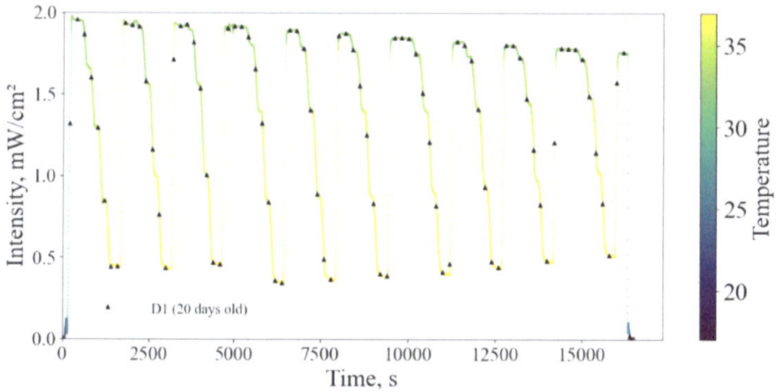

Fig. 5. Variation of the diffracted beam signal with changing temperature over 10 temperature cycles.

5 Conclusions

This study demonstrates the successful fabrication of holographic couplers for fluorescence-based waveguiding, achieving efficient in- and out-coupling of optical signals. The recorded gratings exhibited a spatial frequency of 1720 ± 5 lines/mm with an average diffraction efficiency of 60%, ensuring excitation light coupling through total internal reflection. A secondary holographic optical element enabled selective out-coupling of the fluorescence signal at 700 nm, supporting real-time optical sensing. Additionally, angular and peristrophic multiplexing allowed simultaneous excitation and signal detection, confirming the system's potential for spatial mapping and multiplexed sensing.

The temperature-sensitive gratings exhibited a repeatable Bragg shift response within the clinically relevant range of 33–37 °C, demonstrating their effectiveness as optical transducers for contact-based temperature sensing. A novel photopolymer composition with varied monomer ratio and crosslinker content (25% AA: 75% DA: 25% BA) achieved a fivefold increase in sensitivity, with a normalised intensity change of 18%/°C. The recorded gratings had a spatial frequency of 820 ± 5 lines/mm and an initial diffraction efficiency of 50 ± 10%, ensuring operation within a linear Bragg regime. The laminated structures retained stable performance over 10 cycles, reinforcing their suitability for long-term, non-invasive wound monitoring.

References

1. C. K. Sen, "Human Wound and Its Burden: Updated 2022 Compendium of Estimates," Adv. Wound Care 12, 657–670 (2023).
2. G. Jf, F. Gw, and V. P, "Cohort study evaluating the burden of wounds to the UK's National Health Service in 2017/2018: update from 2012/2013," BMJ Open 10, (2020).
3. G. Gc, W. S, B. Y, and L. Mt, "Wound repair and regeneration," Nature 453, (2008).
4. D. Chakraborty, R. Georgiev, S. Aspell, V. Toal, I. Naydenova, D. Cody, and S. Martin, "Modelling and Design of Holographic Optical Elements for Beam-Coupling Applications for a Range of Incident Beam Angles," Photonics 9, 936 (2022).
5. P. Stoeva, T. Mikulchyk, B. Rogers, M. Oubaha, S. Martin, D. Cody, A. Ferrara, G. Coppola, and I. Naydenova, "Development of holographic optical elements for use in wound monitoring," in Holography: Advances and Modern Trends VIII, A. Fimia and M. Hrabovský, eds. (SPIE, 2023), p. 4.
6. G. Antony, D. Cody, and I. Naydenova, "Theoretical study of slanted holographic gratings for application in sensing," Asian J. Phys. 32, 291 (2023).
7. P. Stoeva, D. Cody, S. Martin, and I. Naydenova, "Holographic Photopolymer Gratings: A Step toward Wound Temperature Sensing," ACS Appl. Opt. Mater. (2025).
8. M. Fierheller and R. G. Sibbald, "A Clinical Investigation into the Relationship between Increased Periwound Skin Temperature and Local Wound Infection in Patients with Chronic Leg Ulcers," Adv. Skin Wound Care 23, 369 (2010).
9. T. Mikulchyk, M. Oubaha, A. Kaworek, B. Duffy, M. Lunzer, A. Ovsianikov, S. Gul, I. Naydenova, and D. Cody, "Synthesis of Fast Curing, Water-Resistant and Photopolymerizable Glass for Recording of Holographic Structures by One- and Two-Photon Lithography," Articles (2022).
10. T. Mikulchyk, P. Stoeva, A. Kaworek, M. Oubaha, B. Rogers, S. Martin, D. Cody, and I. Naydenova, "Characterisation of Holographic Recording in Environmentally Stable Photopolymerisable Glass," Appl. Sci. 12, 5969 (2022).
11. I. Naydenova, "Chapter 7 - Holographic Sensors," in Optical Holography, P.-A. Blanche, ed. (Elsevier, 2020), pp. 165–190.

12. A. Cubells-Gómez, M. I. Lucío, M.-J. Bañuls, and Á. Maquieira, "Holographic Surface Relief Diffraction gratings made of hydrogels for direct label-free biosensing of IgGs," Talanta 126563 (2024).
13. G. Antony, D. Cody, S. Mintova, and I. Naydenova, "Theoretical and Experimental Investigation of Slanted Volume Transmission Gratings for Holographic Gas Sensing," IEEE Sens. J. 1–1 (2024).
14. F. K. Sartain, X. Yang, and C. R. Lowe, "Holographic Lactate Sensor," Anal. Chem. 78, 5664–5670 (2006).

Part V
XR (AR, MR, VR) and Metaverse LDC5

3D Displays Using Interleaved Fresnel Lenses

Hideki Kakeya[0000-0003-3788-9133] and Hiroto Omori

University of Tsukuba, Tsukuba Ibaraki 3058573, Japan
kake@iit.tsukuba.ac.jp

Abstract. We propose a 3D display system based on coarse integral imaging with a seamlessly aligned Fresnel lens array composed of interleaved elemental lenses. By using elemental lenses with smaller widths, the parallax between adjacent elemental images is reduced, resulting in improved image continuity. Additionally, the smaller parallax leads to clearer images with minimal blurring, even when the images are blended by the interleaved Fresnel lens array. To address the issue of a narrow viewing zone due to the smaller elemental lenses, eye-tracking technology is used to follow the motion of the viewer, enabling the generation of light space with high spatial resolution. Real-time generation of elemental images is achieved on iPhone and iPad platforms. Uniformity of image is improved by using an EL display in place of an LCD display.

Keywords: Integral Imaging, Resolution, Lens Array

1 Introduction

Integral imaging [1] realizes an autostereoscopic display by reproducing a light-ray space with a fly-eye lens. Unlike the autostereoscopic displays using lenticular lenses, integral imaging can present parallax in both horizontal and vertical directions.

To reproduce a high-density light ray space with integral imaging, resolution of the elemental images has to be sufficiently high. Therefore, it is a challenge to present high-resolution 3D images at positions far from the screen. One way to address this issue is to use a lens array with larger elemental lenses instead of a microlens array [2-5]. Coarse integral imaging (CII) and coarse integral volumetric imaging (CIVI) [6-8] generate either real or virtual images with corrected optical and color aberration in the air, which enables natural and clear 3D images with little vergence-accommodation conflict away from the screen.

However, the use of a coarse lens array leads to a noticeable image discontinuity at the seams of elemental lenses. To address this issue, an interleaved Fresnel lens array, where the elemental prisms of adjacent elemental lenses are interleaved at the boundary, has been proposed to improve image continuity [9].

To further improve image continuity, a lens array with interleaved grooves across the entire lens has been proposed [10]. Applying this interleaved Fresnel lens array to CII enables blending of adjacent elemental images, thereby improving image continuity [11,12]. However, when presenting 3D images with deep depth, the increased parallax between adjacent elemental images leads to image blur.

S. Kurimura and T. Yagi (eds.), *Proceedings of Laser Display and Lighting Conference 2025*, Springer Proceedings in Physics 429,
https://doi.org/10.1007/978-981-96-9782-3_28

A CII display using a fine interleaved Fresnel lens array has been proposed to address these issues. By reducing the width of the elemental lenses that compose the lens array, parallax between adjacent elemental images is decreased linearly, thereby improving image continuity. Additionally, even when adjacent elemental images are blended by the interleaved lenses, smaller parallax between them is expected to result in less blurring.

Fine pitch of the lens array, however, narrows the viewing zone. To address this issue, research has been conducted on methods to expand the viewing zone using eye-tracking technology [13]. This method solves the problem of viewing zone limitation caused by the smaller width of the elemental lenses.

To generate elemental images that follow the position of the observer, real-time rendering is required. As the width of the elemental lenses decreases, the number of views to be rendered increases, which requires significant computational resources. The use of single-pass multi-view rendering [14] and meshlet culling [15] has been proposed to overcome this problem.

In this paper, we present recent advancements in coarse integral imaging with interleaved lens array. We have created a large seamless metal mold of interleaved Fresnel lens array to realize a seamless 3D image with a large screen size. Real-time generation of elemental images is achieved on iPhone and iPad platforms. The uniformity of image is improved by using an EL display in place of an LCD display.

2 Coarse integral imaging

Conventional integral imaging uses a microlens array to realize a multi-view display. The number of pixels covered by each elemental lens matches the number of views, and one pixel is observed through each elemental lens. As a result, the observer perceives each elemental lens as a single pixel.The panel and the lens array are arranged so that the distance between them may be equal to the focal length of the elemental lenses to collimate light rays.

When the resolution of elemental images is high, dense light field is generated, which is hard to attain with a conventional electronic display. Thus, the resolution of conventional integral imaging significantly decreases as the distance from the display increases, making it difficult to present high-resolution 3D images at a position far from the display.

CII uses a lens array composed of larger elemental lenses. Figure 1 illustrates the principle of CII. Since the number of pixels covered by each elemental lens is greater than the number of views, multiple pixels are observed as real or virtual images through each elemental lens, which enables presentation of natural 3D images far from the display hardware.

However, using a coarse lens array increases the parallax between adjacent elemental images, resulting in decreased image continuity. One way to solve this problem is to use an interleaved Fresnel lens array in place of a conventional lens array.

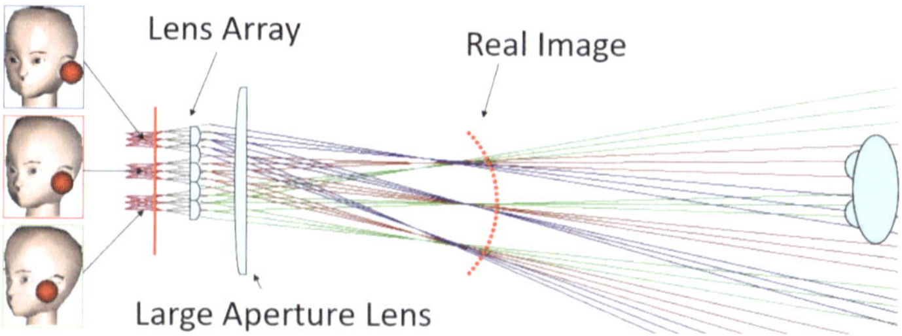

Fig. 1. Coarse integral imaging.

3 Interleaved Fresnel lens array

In the interleaved Fresnel lens, elemental prisms of adjacent elemental lenses with different optical centers are interleaved, as shown in Fig. 2. The design of the elemental prisms in the interleaved Fresnel lens array used here is shown in Fig. 3. This lens array is designed so that the paired adjacent elemental prisms may have the same height. Consequently, the width ratio of the adjacent elemental prisms gradually changes, which realizes a natural shift of elemental image blending.

Let the tilt angles of paired adjacent elemental prisms be α_l and α_r, and the height of the elemental prisms be h. The widths w_l and w_r of the paired elemental prisms are given by

$$w_l = \frac{h}{\tan \alpha_l}, \tag{1}$$

$$w_r = \frac{h}{\tan \alpha_r}. \tag{2}$$

When the total width of the adjacent elemental prisms is $w = w_l + w_r$, the height h is given by

$$h = \frac{w \tan \alpha_l \tan \alpha_r}{\tan \alpha_l + \tan \alpha_r}. \tag{3}$$

Substituting equation (3) into equations (1) and (2) gives

$$w_l = \frac{w \tan \alpha_r}{\tan \alpha_l + \tan \alpha_r}, \tag{4}$$

$$w_r = \frac{w \tan \alpha_l}{\tan \alpha_l + \tan \alpha_r}, \tag{5}$$

$$\frac{w_l}{w_r} = \frac{\tan \alpha_r}{\tan \alpha_l}. \tag{6}$$

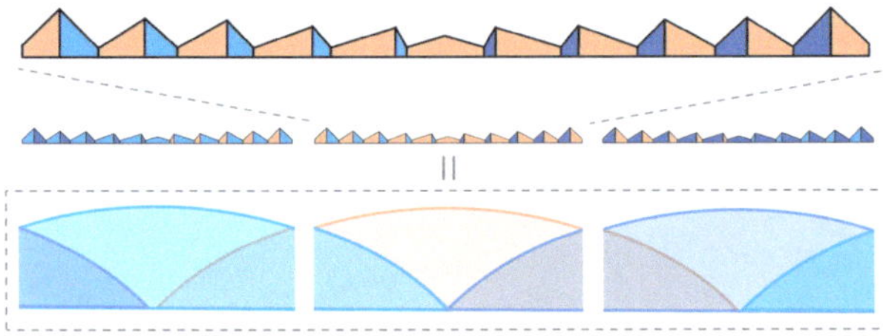

Fig. 2. Interleaved Fresnel lens.

An interleaved Fresnel lens array is realized by overlapping linear Fresnel lenses with interleaved grooves orthogonally. Figure 4 shows the design of the linear Fresnel lenses used in this study. The width and the focal length of the elemental lens are 4.8 mm and 15 mm respectively. The total width of two adjacent elemental prisms w is 0.24 mm.

When a pair of the linear interleaved Fresnel lenses are stacked together orthogonally, an interleaved convex lens array is generated, which can be used as the lens array for integral imaging with blending effect.

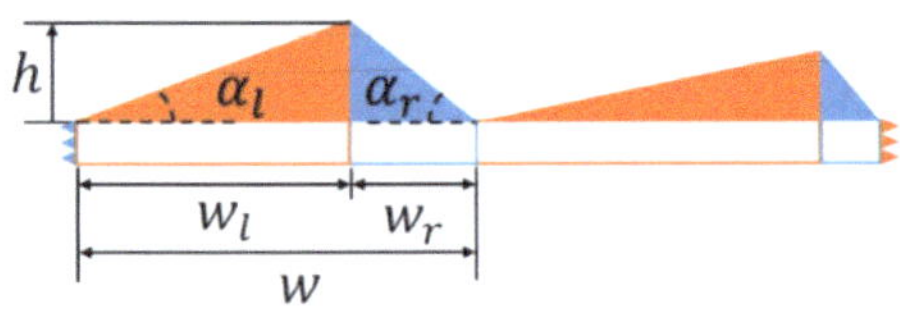

Fig. 3. Elemental prisms of the interleaved Fresnel lens.

Fig. 4. Design of the interleaved elemental lens.

4 Coarse integral imaging with interleaved lenses

The viewing zone of the CII is defined by the area where the viewing zones of all the elemental images overlap, as shown in Fig. 5. The size of elemental images can be adjusted as the distance between the observer and the lens array changes. If the width of the elemental lenses is d and the distance from the observer to the lens array is L, the optimized size of the elemental images p is given by

$$p = \frac{L + g}{L} d . \tag{8}$$

As shown in Fig. 6, it is possible to keep on providing proper 3D image for the observer by shifting the positions and updating the contents of elemental images, which is enabled by utilizing eye-tracking technology and real-time rendering to follow the position of the observer.

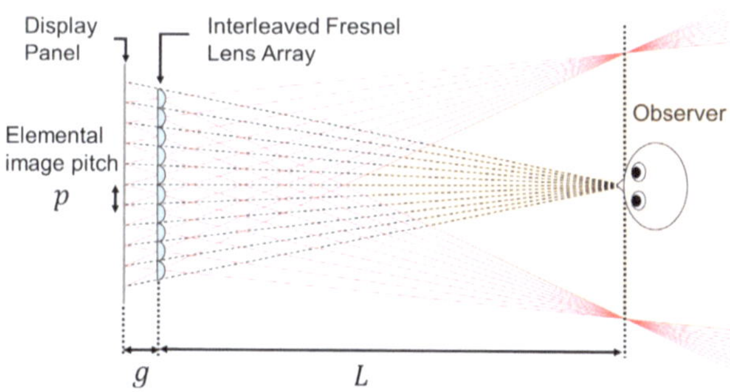

Fig.5. Viewing zone of CII.

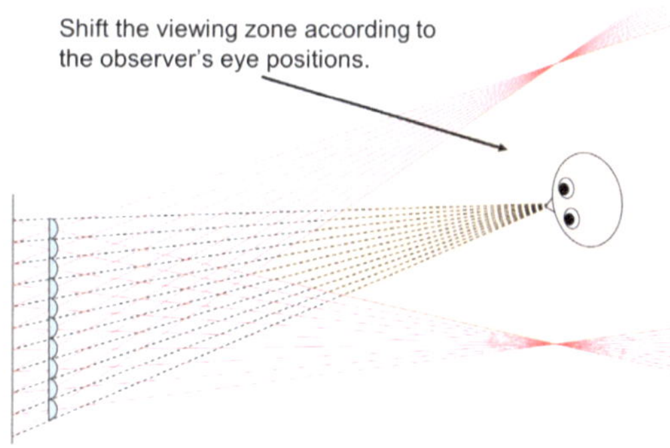

Fig. 6. Viewing zone of CII when face tracking is applied to follow the observer's motion.

5 Recent advancement

The conventional interleaved Fresnel lens was small and many pieces were patched together to realize a large screen. Therefore, the observed image included seams of lens array pieces, which degrades the impression of the image especially when a real image is generated in front of the lens and the screen. To address this issue, we made a large seamless metal mold of interleaved Fresnel lens array to realize a seamless 3D image with a large screen size. The size of the lens array was 345.6 mm × 345.6 mm, comprising 72 elemental lenses. The picture of a lens array generated with a metal mold is shown in Fig.7, where two lens arrays are stacked together to generate a convex lens array composed of 72 × 72 elemental lenses.

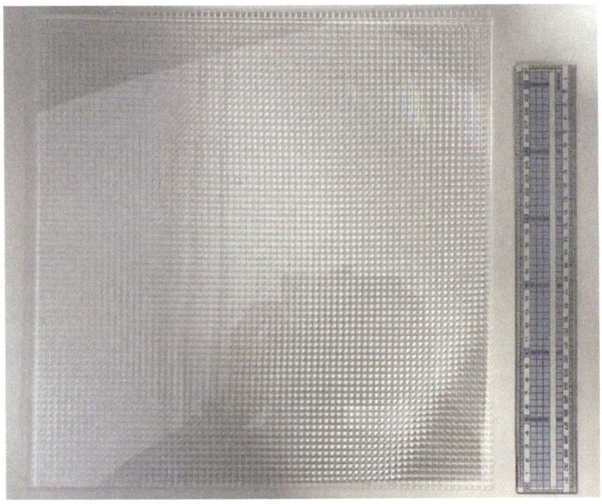

Fig. 7. Picture of a large lens array.

Figure 8 shows a CII prototype system and the generated image observed from top, left, right, and bottom positions. Here the display panel and the lens array were arranged to keep 19 mm interval, which is longer than the focal distance of the elemental lenses, to generate a real image 71 mm in front of the lens array. A camera was installed on top of the system to track the position of an observer.

By using an interleaved Fresnel lens array, the adjacent elemental images are blended, resulting in improved image continuity and clarity. Since the parallax between adjacent elemental images is small, a clear image is attained even when the images are blended.

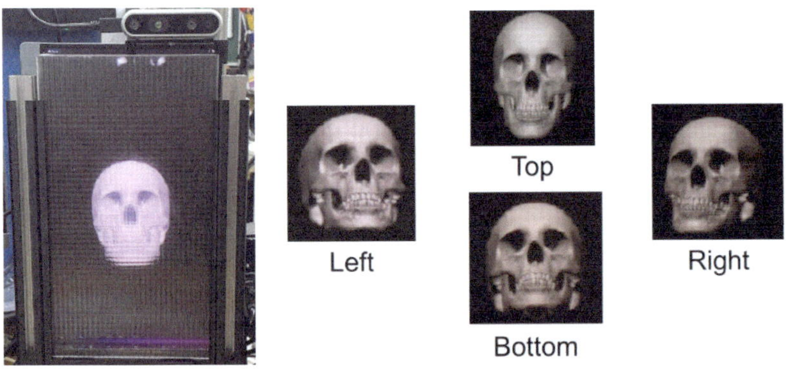

Fig. 8. a CII prototype system and the generated image observed from top, left, right, and bottom positions.

Real-time generation of elemental images is achieved on iPad (Fig. 9) and iPhone (Fig. 10) platforms. The lenses can be attached to iPad or iPhone to realize a portable 3D display. The real-time calculation on these platforms is realized with single-pass multi-view rendering [14].

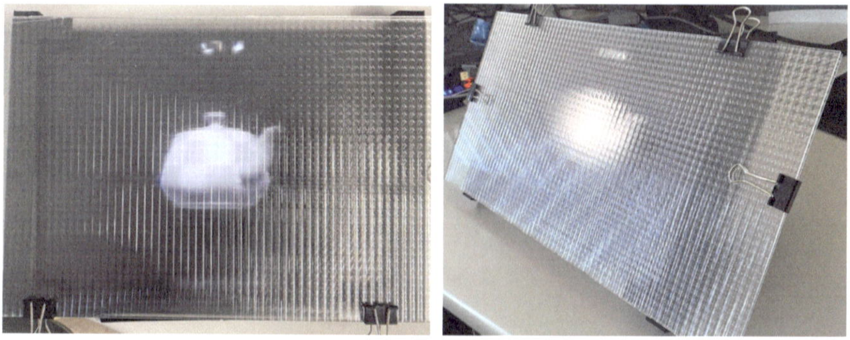

Fig. 9. CII with iPad platform.

Fig. 10. CII with iPhone platform.

Though the interleaved lens blends the light from two neighboring elemental images, the light passing through the lens is not perfectly uniform because the light emitted from the display surface weakens as the angle deviates from the upright position, especially when an LCD panel is used for the display.

Figure 11 shows the nonuniformity of light when the interleaved lens array is applied to an iPad with an LC display. The nonuniformity increases as the viewing angle becomes wider.

Figure 12 shows the image of the screen when the interleaved lens array is applied to an iPad with an organic EL display. Because of the self-emitting light with high luminance across a wider angle, the uniformity of luminance is enhanced with the use of an EL display.

Fig. 11. Uniformity of light given by CII using iPad with an LC display.

Fig. 12. Uniformity of light given by CII using iPad with an EL display.

6 Summary

Recent advancements of coarse integral imaging with interleaved lens array are reported. A large seamless metal mold of interleaved Fresnel lens array wad manufactured to realize a seamless 3D image with a large size. Real-time generation of elemental images is achieved on iPhone and iPad platforms using single-pass multi-image rendering. Uniformity of image is improved by using an EL display in place of an LCD display.

Further improvement of image uniformity is expected by using a laser display behind the lens array or by compensation of image nonuniformity with an luminance adjustment sheet on top of the lens array or image correction in each elemental image on the display panel.

Acknowledgements

This research is partially supported by the Grant-in-Aid for Scientific Research, JSPS, Japan, Grant number: 22H03624 and by JST CREST Grant Number: JPMJCR18A2.

References

1. Lippmann, G.: La photographie integrale. C. R. Acad. Sci. vol. 146, 446–451 (1908).
2. Lee, B., Jung, S., Min, S.-W., and Park, J.-H.: Three-dimensional display by use of integral photography with dynamically variable image planes. Opt. Lett. 26(19), 1481–1482 (2001).
3. Park, J.-H., Jung, S., Choi, H., and Lee, B.: Integral imaging with multiple image planes using a uniaxial crystal plate. Opt. Express 11(16), 1862–1875 (2003).
4. Kim, Y., Park, J.-H., Choi, H., Kim, J., Cho, S.-W., and Lee, B.: Depth-enhanced three-dimensional integral imaging by use of multilayered display devices. Appl. Opt. 45(18), 4334–4343 (2006).
5. Kim, Y., Choi, H., Kim, J., Cho, S.-W., Kim, Y., Park, G., and Lee, B.: Depth-enhanced integral imaging display system with electrically variable image planes using polymer-dispersed liquid-crystal layers. Appl. Opt. 46(18), 3766–3773 (2007).
6. Kakeya, H.: Coarse integral imaging and its applications. Proc. SPIE 6803, 680317 (2008).
7. Kakeya, H.: Improving image quality of coarse integral volumetric display. Proc. SPIE 7237, 723726 (2009).
8. Kakeya, H.: Realization of undistorted volumetric multiview image with multilayered integral imaging. Opt. Express 19(21), 20395–20404 (2011).
9. Kakeya, H. and Sawada, S.: Reduction of image discontinuity in coarse integral volumetric imaging. Opt. Lett. 40(23), 5698–5701 (2015).
10. Borjigin, G. and Kakeya, H.: A backlight system using a novel interleaved Fresnel lens array that attains a uniform luminance and two-dimensional directional light control. Opt. Lett. 47(2), 301–304 (2022).
11. Kakeya, H. and Borjigin, G.: Realization of smooth motion parallax by coarse integral imaging with interleaved elemental Fresnel lenses. Proc. IDW, 529–532 (2022).
12. Borjigin, G. and Kakeya, H.: Coarse integral imaging displays with interleaved Fresnel lenses. SID Digest of Technical Papers 54, 518-521 (2023).
13. Park, G., Jung, J. H., Hong, K., Kim, Y., Kim, Y. H., Min, S. W., and Lee, B.: Multiviewer tracking integral imaging system and its viewing zone analysis. Opt. Express 17(20), 17895–17908 (2009).
14. Omori, H. and Kakeya, H.: Coarse integral imaging with eye tracking using fine interleaved Fresnel lens arrays for enhanced viewing angles and image continuity. ITE Trans. MTA, 13(1), 2–7 (2025).
15. Omori, H. and Kakeya, H.: Meshlet-based efficient elemental image generation for coarse integral imaging. Proc. IDW, vol. 31 742-745 (2024).

From Design to Display: Tackling Manufacturing Challenges in AR Waveguides

Juha Pietarinen[1] and Joonas Pylväinenr[1]

[1] Optofidelity Oy, Metallimiehenkuja 10, 02150 Espoo, Finland

www.Optofidelity.com
juha.pietarinen@optofidelity.com

Abstract. This discussion explores Diffractive Waveguides in AR smart glasses, emphasizing their impact on visual quality and user experience. It highlights the importance of precision in manufacturing and rigorous quality assurance, which are key foundations for optimizing processes, minimizing errors, and delivering high-performing waveguide displays for immersive augmented reality experiences. As AR becomes more widely used, the intended and expected experiences will become more complex, as the below Commercialization Timeline by Ericsson describes. Currently many smart glasses are offering simple text-based displays in single color. Over the next five years, the user experience is projected to become far more complex.

Keywords: Diffractive Waveguide, AR smart glass, Quality Control, User Experience, Process Optimization.

1 Introduction

AR smart glasses represent a cutting-edge technology, seamlessly integrating digital overlays with the physical world to deliver transformative user experiences. The optical display systems in these glasses rely on sophisticated technologies. At their core, a display panel produces digital visuals, which are projected via compact optics and transmitted to the user's eyes using transparent optical waveguides [3]. Among the various available technologies, diffractive waveguides stand out for their ability to deliver superior image quality.

This paper examines the critical role of manufacturing accuracy and stringent quality control in the performance of diffractive waveguides. Addressing production challenges and reducing errors ensure exceptional AR experiences.

At Meta Connect 2024, Mark Zuckerberg revealed that the Orion project faced a mere 10% chance of success, underscoring the immense difficulties of scaling AR glasses for mass production [1]. Developing and integrating technologies like diffractive,

S. Kurimura and T. Yagi (eds.), *Proceedings of Laser Display and Lighting Conference 2025*, Springer Proceedings in Physics 429,
https://doi.org/10.1007/978-981-96-9782-3_29

holographic, and reflective waveguides into compact, efficient designs has posed significant challenges.

Currently, diffractive waveguides dominate as the preferred technology for delivering vivid, transparent displays that blend virtual imagery with real-world visuals. Many AR products today leverage this technology, supported by advancements in nanoimprint lithography (NIL), a process ideal for mass-producing the intricate grating structures required for waveguides [4-7]. Microsoft's Alex Kipman likened the precision needed in this field to positioning a coin correctly from a kilometer away within a few arc seconds [2].

1.1 The Complexity of Waveguide Manufacturing

Nanoimprint lithography has become a go-to method for producing high-volume diffraction gratings. However, manufacturing involves numerous intricate processes, each introducing variables that must be meticulously managed. From material properties to environmental conditions, every parameter can affect performance.

Even small errors in diffraction angles can lead to issues like uneven brightness, blurry visuals, or ghost images. Achieving top-tier optical quality requires tight control over critical-to-quality (CTQ) factors, which customers value most, such as sharpness, contrast, and uniformity.

For instance, image clarity—measured by the Modulation Transfer Function (MTF)—is influenced by everything from initial design to the properties of the materials and even minor cosmetic defects. This underscores the importance of identifying and managing critical-to-function (CTF) parameters during production.

MTF depends on both front-end and back-end processes. In back-end steps, projector performance and the alignment of stacked waveguides for expanded fields of view are vital. Cosmetic defects like scratches or delamination also diminish MTF and overall image quality.

On the front end, the precise imprinting of grating dimensions—such as depth, period, and orientation—is crucial. Material properties, like resin viscosity and refractive index, alongside substrate characteristics, must also meet strict specifications. Proper control of equipment parameters like curing time and speed further ensures consistency.

2 The Path Forward: Structured Manufacturing Solutions

Addressing these challenges requires a strategic approach that begins with stabilizing production lines and extends to minimizing defects and improving yields.

Early detection and resolution of defects are pivotal. Identifying issues at the start of production is not only cost-effective but also critical for maintaining quality standards.

Investing in robust metrology systems ensures that key parameters are consistently monitored, enabling swift interventions when anomalies arise.

The well-known 4M framework—Man, Material, Machine, and Method—provides a structured approach to optimizing waveguide production.

Man: Skilled technicians and engineers are essential for monitoring processes and ensuring quality.

Material: High-quality raw materials, including substrates and resins, must meet rigorous standards.

Machine: Equipment like NIL systems requires precise calibration to produce consistent results.

Methods: Lean methodologies, such as Six Sigma, guide process improvements and defect minimization.

By systematically managing these factors, manufacturers can achieve consistent quality and reduce the likelihood of defects.

3 Conclusion: Mastering Complexity for AR Innovation

Producing AR glasses with high-performance waveguides is a multifaceted challenge requiring attention to every detail—from material properties to process control. Success hinges on adopting a holistic approach that integrates robust quality assurance, early defect detection, and ongoing optimization.

By understanding and tracking the most critical parameters, manufacturers can enhance yields, lower costs, and deliver the superior image quality that defines next-generation AR experiences. With these measures in place, the industry is well-positioned to meet the growing demand for advanced smart glasses and exceed user expectations.

References

1. Meta, "Meta Connect," 25 September 2024. [Online]. Available: https://developers.facebook.com/m/meta-connect-developer-sessions/connect-keynote/.
2. Microsoft, "ETH Zürich," 3 October 2019. [Online]. Available: https://video.ethz.ch/speakers/global-lecture/2019/61d6198e-bf0d-4970-962a-a56e70482fce.html.
3. L. e. a. Barbato, "Augmented Reality Using Diffractive Waveguides: Challenges and Opportunities.," Applied Optics, OSA Publishing, [Online]. Available: https://www.osapublishing.org/ao/abstract.cfm?uri=ao-58-34-9566.
4. Q. e. a. Zhang, "Advanced Fabrication Techniques for Diffractive Waveguides in AR Displays.," Light: Science & Applications, Nature Publishing, [Online]. Available: https://www.nature.com/articles/s41377-023-01371-4.
5. D. Cheng, Q. Wang, Y. Liu, H. Chen, D. Ni, X. Wang, C. Yao, Q. Hou, W. Hou, G. Luo, and Y. Wang, "Design and Manufacture AR Head-Mounted Displays: A Review and Outlook," Light: Advanced Manufacturing, vol. vol. 2, p. p. 24, 2021.

6. B. C. Kress and A. Shirai, Optical Architectures for Augmented-, Virtual-, and Mixed-Reality Headsets: Fundamentals, Theory, and Applications for Optical Design, Bellingham, WA: SPIE Press, 2020.
7. B. Kress and I. Chatterjee, "Waveguide Combiners for Mixed Reality Headsets: A Nanophotonics Design Perspective," Nanophotonics, vol. 1, 2020.
8. D. Wagner, "Why Making Good AR Displays is So Hard," [Online]. Available: https://www.linkedin.com/pulse/why-making-good-ar-displays-so-hard-daniel-wagner/.

Projection Mapping Technologies for Seamless Spatial Augmentation

Daisuke Iwai[1[0000-0002-3493-5635]]

[1] The University of Osaka, Toyonaka, Osaka 5608531, Japan
daisuke.iwai.es@osaka-u.ac.jp

Abstract. Projection mapping (PM) enables multiple users to experience virtual and augmented reality simultaneously with the naked eyes, making it valuable beyond traditional art and entertainment applications. It has been applied in various spatial augmentation domains, such as surgical assistance, object search, industrial design, remote communication, and human augmentation. However, achieving seamless spatial augmentation remains challenging due to PM's limitations in bright environments and shadow interference during user interactions. Recent technical advancements in the PM research field have addressed these issues. Specifically, it explains a technique applying a multi-projector system that selectively illuminates areas other than the projection target, maintaining high contrast in a bright room. Light-field projectors simulate room lighting, enabling realistic shadow behavior. Shadow removal is achieved using large-aperture optics like micro-mirror array plates and Fresnel lenses. Future projection mapping systems must focus on simpler calibration methods and energy-efficient laser-based projectors to overcome remaining challenges. This talk introduces these technologies for seamless spatial augmentation.

Keywords: Augmented Reality, Display Technology, Projection Mapping.

1 Introduction

Projection mapping (PM), which overlays visual content onto real-world objects in the surrounding environment, offers a distinct advantage over other display technologies such as head-mounted displays (HMDs) or hand-held flat-panel displays (e.g., smartphones). This advantage lies in its ability to allow multiple users to simultaneously experience virtual reality (VR) and augmented reality (AR) with the naked eyes. Leveraging this characteristic, PM has been explored not only in traditional art and entertainment applications but also in various other spatial augmentation domains, including surgical assistance [1], object search support [2,3,4], industrial design assistance [5,6], remote communication [7,8], makeup [9,10,11], and human augmentation (hand augmentation [12,13,14,15], vision augmentation [16,17], and extended body [18,19]).

Unlike art and entertainment applications designed to create unusual experiences, these spatial augmentation applications are intended for use in everyday life. However,

S. Kurimura and T. Yagi (eds.), *Proceedings of Laser Display and Lighting Conference 2025*, Springer Proceedings in Physics 429,
https://doi.org/10.1007/978-981-96-9782-3_30

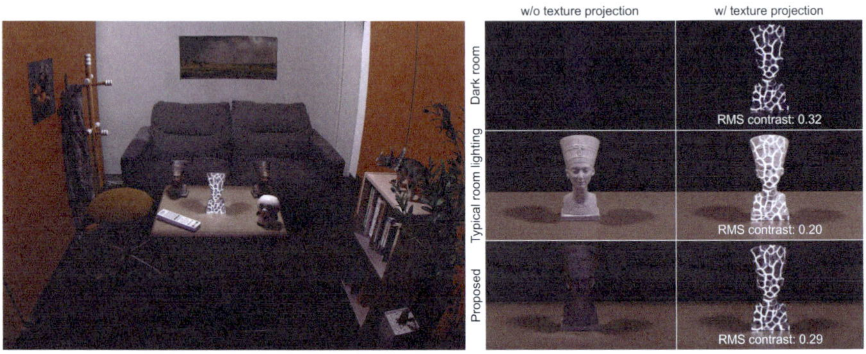

Fig. 1. The PM system reduces contrast loss under environmental lighting [20]. The left image shows texture projection in a well-lit room using their technique. A crack pattern was projected onto the statue at the desk's center. The right images compare the statue without and with texture projection under three lighting conditions: a dark room, typical LED room lighting, and their proposed projector-based lighting. The results highlight enhanced contrast and lighting reproduction.

the technical limitations of PM pose challenges in achieving seamless spatial augmentation. Specifically, in bright environments, the contrast of projected content decreases, which is why PM is typically performed in dark environments. It is undesirable to require darkening the room every time a spatial augmentation application is used. Additionally, when a user steps between the projector and the target surface, shadows are cast on the projection, making it difficult to view the necessary information. This issue becomes critical in interactive spatial augmentation applications, where users manipulate projected content by touching it on the target surface. This invited talk will introduce the latest technologies that address these limitations to realize seamless spatial augmentation.

2 Towards Seamless Spatial Augmentation

A PM system capable of maintaining contrast in a bright environment can be realized by using projectors as lighting that selectively illuminates areas other than the projection target. Takeuchi et al. proposed a technique that uses multiple projectors to reproduce the illuminance of room lighting while avoiding illumination of the projection target [20] (Fig. 1). Yasui et al. developed a system employing a projector and a 2D lens array to reproduce the light field of room lighting, enabling shadows caused by user obstruction to appear similar to those produced by usual room lighting [21]. These technologies have demonstrated that PM with high contrast is feasible in bright rooms. However, these techniques require geometric and photometric calibration of multiple projectors, which involves complex processes requiring expertise to achieve the necessary precision. Recently, a technique utilizing neural networks to perform these calibration processes end-to-end has been developed [22], but further advancements in simpler and more accurate calibration techniques are still needed.

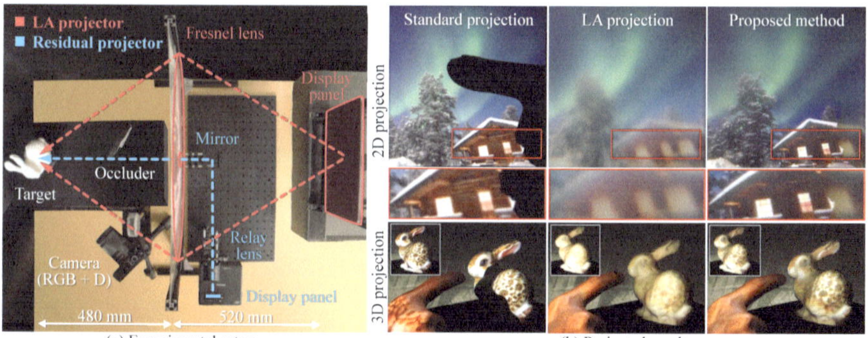

(a) Experimental setup (b) Projected results

Fig. 2. Shadow suppression by the multi-aperture coaxial projector [26]: (a) Prototype system with a large-aperture (LA) projector using a Fresnel lens and a standard projector. (b) Results showing standard projection (with shadows), LA projection (shadow suppressed but low resolution), and their proposed method (shadow and quality issues resolved).

Regarding shadow removal, techniques have been proposed to compensate for occluded light from one projector by using light from another projector placed in a spatially distributed configuration [23]. However, as mentioned above, the calibration of multiple projectors is not a trivial task. To address this, new shadow removal technologies have been developed that do not require the calibration by using optical systems capable of projecting images from a wide range of directions. Hiratani et al. achieved large-aperture projection by utilizing an optical element called a micro-mirror array plate (MMAP), commonly used in aerial displays [24,25]. Although this method significantly reduced shadow generation, the MMAP requires the system to be very large. To address this issue, Kusuyama et al. proposed a technique that employs a Fresnel lens for large-aperture projection [26] (Fig. 2). While the use of Fresnel lenses can cause significant image quality degradation in the projected results, they combined the Fresnel lens-based large aperture projector with a conventional projector and aligned them coaxially. Thereby, they achieved both shadow suppression and minimal image quality degradation.

3 Conclusion

This invited talk introduced recent PM technologies that have been developed to realize seamless spatial augmentation. The *seamlessness* is a new grand challenge in PM, and thus, there are still many unsolved technical issues, such as simple projector calibration and high-fidelity appearance reproduction. To this end, a multi-projection system is essential, and thus, projectors applying laser-based light sources requiring less energy will become more demanding. Laser display technologies will be one of the key components of future PM systems. Interested readers may refer to the latest survey article in PM [27].

References

1. Nishino, H., Hatano, E., Seo, S., Nitta, T., Saito, T., Nakamura, M., Hattori, K., Takatani, M., Fuji, H., Taura, K., Uemoto S.: Real-time Navigation for Liver Surgery Using Projection Mapping with Indocyanine Green Fluorescence: Development of the Novel Medical Imaging Projection System. Annals of Surgery 267(6), 1134-1140 (2018).
2. Iwai, D., Sato, K.: Document search support by making physical documents transparent in projection-based mixed reality. Virtual Reality 15, 147-160 (2011).
3. Matsushita K., Iwai, D., Sato, K.: Interactive bookshelf surface for in situ book searching and storing support. In: Proceedings of Augmented Human International Conference, Article 2, ACM (2011).
4. Kitajima, Y., Iwai, D., Sato, K.: Simultaneous Projection and Positioning of Laser Projector Pixels. IEEE Transactions on Visualization and Computer Graphics 23(11), 2419-2429 (2017).
5. Takezawa, T., Iwai, D., Sato, K., Hara, T., Takeda, T., Murase, K.: Material Surface Reproduction and Perceptual Deformation with Projection Mapping for Car Interior Design. In: Proceedings of IEEE Conference on Virtual Reality and 3D User Interfaces, pp. 251-258, IEEE (2019).
6. Cascini, G., O'Hare, J., Dekoninck, E., Becattini, N., Boujut, J.-F., Guefrache, F. B., Carli, I., Caruso, G., Giunta, L., Morosi, F.: Exploring the use of AR technology for co-creative product and packaging design. Computers in Industry 123, Article 103308 (2020).
7. Pejsa, T., Kantor, J., Benko, H., Ofek, E., Wilson, A.: Room2Room: Enabling Life-Size Telepresence in a Projected Augmented Reality Environment. In: Proceedings of the 19th ACM Conference on Computer-Supported Cooperative Work & Social Computing, pp. 1716–1725, ACM (2016).
8. Iwai, D., Matsukage, R., Aoyama, S., Kikukawa, T., Sato, K.: Geometrically Consistent Projection-based Tabletop Sharing for Remote Collaboration. IEEE Access 6, 6293-6302 (2018)
9. Bermano, A. H., Billeter, M., Iwai, D., Grundhöfer, A.: Makeup Lamps: Live Augmentation of Human Faces via Projection. Computer Graphics Forum 36(2), 311-323 (2017).
10. Siegl, C., Lange, V., Stamminger, M., Bauer, F., Thies, J.: FaceForge: "Markerless Non-Rigid Face Multi-Projection Mapping. IEEE Transactions on Visualization and Computer Graphics 23(11), 2440-2446 (2017).
11. Peng, H., Nishida, S., Watanabe, Y.: Studying User Perceptible Misalignment in Simulated Dynamic Facial Projection Mapping. In: Proceedings of IEEE International Symposium on Mixed and Augmented Reality, pp. 493-502, IEEE (2023).
12. Mistry P., Maes, P.: SixthSense: a wearable gestural interface. In: Proceedings of ACM SIGGRAPH ASIA 2009 Art Gallery & Emerging Technologies: Adaptation, Article 85, ACM (2009).
13. Harrison, C., Benko, H., Wilson, A. D.: OmniTouch: wearable multitouch interaction everywhere. In: Proceedings of the 24th annual ACM symposium on User interface software and technology, pp. 441-450, ACM (2011).
14. Sodhi, R., Benko, H., Wilson, A.: LightGuide: projected visualizations for hand movement guidance. In: Proceedings of the SIGCHI Conference on Human Factors in Computing Sytems, pp. 179-188, ACM (2012).
15. Erel, Y., Mordenfeld, O. K., Iwai, D., Sato, K., Bermano, A. H.: Casper DPM: Cascaded Perceptual Dynamic Projection Mapping onto Hands. In: Proceedings of ACM SIGGRAPH Asia 2024 Conference Papers, Article No. 137, ACM (2024).

16. Ueda, T., Iwai, D., Hiraki, T., Sato, K.: Illuminated Focus: Vision Augmentation using Spatial Defocusing via Focal Sweep Eyeglasses and High-Speed Projector. IEEE Transactions on Visualization and Computer Graphics 26(5), 2051-2061 (2020).
17. Ueda, T., Iwai, D., Sato, K.: IlluminatedZoom: spatially varying magnified vision using periodically zooming eyeglasses and a high-speed projector. Optics Express 29, 16377-16395 (2021).
18. Ueda, Y., Asai, Y., Enomoto, R., Wang, K., Iwai, D., Sato, K.: Body cyberization by spatial augmented reality for reaching unreachable world. In: Proceedings of the 8th Augmented Human International Conference, Article 19, ACM (2017).
19. Sato, Y., Hiraki, T., Tanabe, N., Matsukura, H., Iwai, D., Sato, K.: Modifying Texture Perception With Pseudo-Haptic Feedback for a Projected Virtual Hand Interface. IEEE Access 8, 120473-120488, (2020).
20. Takeuchi, M., Kusuyama, H., Iwai, D., Sato, K.: Projection Mapping under Environmental Lighting by Replacing Room Lights with Heterogeneous Projectors. IEEE Transactions on Visualization and Computer Graphics 30(5), 2151-2161 (2024).
21. Yasui, M., Iwataki, R., Ishikawa M., Watanabe, Y.: Projection Mapping with a Brightly Lit Surrounding Using a Mixed Light Field Approach. IEEE Transactions on Visualization and Computer Graphics 30(5), 2217-2227 (2024).
22. Erel, Y., Kozlovsky-Mordenfeld, O., Iwai, D., Sato, K., Bermano, A. H.: Casper DPM: Cascaded Perceptual Dynamic Projection Mapping onto Hands. In: Proceedings of ACM SIGGRAPH Asia 2024 Conference Papers, Article No. 137, ACM (2024).
23. Nagase, M., Iwai, D., Sato, K.: Dynamic Defocus and Occlusion Compensation of Projected Imagery by Model-Based Optimal Projector Selection in Multi-projection Environment. Virtual Reality 15(2), 119-132, Springer (2011).
24. Hiratani, K., Iwai, D., Punpongsanon, P., Sato, K.: Shadowless Projector: Suppressing Shadows in Projection Mapping with Micro Mirror Array Plate. In: Proceedings of IEEE Conference on Virtual Reality and 3D User Interfaces, pp. 1309-1310, IEEE (2019).
25. Hiratani, K. Iwai, D., Kageyama, Y., Punpongsanon, P., Hiraki, T., Sato, K.: Shadowless Projection Mapping using Retrotransmissive Optics. IEEE Transactions on Visualization and Computer Graphics 29(5), 2280-2290 (2023).
26. Kusuyama, H., Kageyama, Y., Iwai, D., Sato, K.: A Multi-aperture Coaxial Projector Balancing Shadow Suppression and Deblurring. IEEE Transactions on Visualization and Computer Graphics 30(11), 7031-7041 (2024).
27. Iwai, D.: Projection Mapping Technologies: A Review of Current Trends and Future Directions. Proceedings of the Japan Academy, Series B 100(3), 234-251 (2024).

Enhancing Visual Realism with Projector-Camera Feedback and Light-Field Projection

Toshiyuki Amano [0000-0003-4146-1375]

Wakayama University, Sakaedani 930, Wakayama-city 640-8510, Japan
amano@wakayama-u.ac.jp

Abstract. This paper presents advanced Spatial Augmented Reality (SAR) techniques for precise appearance manipulation, addressing the limitations of traditional methods reliant on Lambertian assumptions. Key contributions include Real-Time Appearance Manipulation, Viewing Direction-Dependent Appearance Manipulation (VDDAM), and Adaptive Light-Field Projection. These methods enable dynamic control of material properties, robust BRDF representation, and seamless applications on complex surfaces, with significant implications for product design, optical restoration, immersive exhibits, and digital twins, bridging physical and digital realms without wearable devices.

Keywords: Spatial Augmented Reality, Appearance Manipulation, Light-Field Projection, BRDF Representation, Real-world Metaverse.

1 Introduction

Spatial Augmented Reality (SAR), first proposed in the early 2000s [1], has explored fundamental technology for projecting visual enhancements directly onto physical objects. Early research focused on challenges such as projection blur compensation, dynamic projection onto moving or deforming objects, and other related issues [2–4].

Over the years, SAR has evolved to include appearance manipulation, enabling the transformation and enhancement of material properties [5–7]. Furthermore, SAR enables the enhancement or transformation of object appearances including optical restoration of deteriorated artworks [8], dynamic range extension for low-contrast objects fabricated with 3D printers [9], and visual enhancement of virtual avatars through the projection [10]. However, these methods often rely on Lambertian reflection assumptions, which fail to accurately model complex light behaviors such as specular reflections and structural colors. To address these limitations, our research explores advanced techniques based on projector-camera feedback systems and light-field projection on the non-Lambert reflection surface for optically correct appearance manipulation. Our key contributions include:

- Real-Time Appearance Manipulation,
- Viewing Direction-Dependent Appearance Manipulation (VDDAM), and
- Adaptive BRDF Display using Light Ray Projection.

S. Kurimura and T. Yagi (eds.), *Proceedings of Laser Display and Lighting Conference 2025*, Springer Proceedings in Physics 429,
https://doi.org/10.1007/978-981-96-9782-3_31

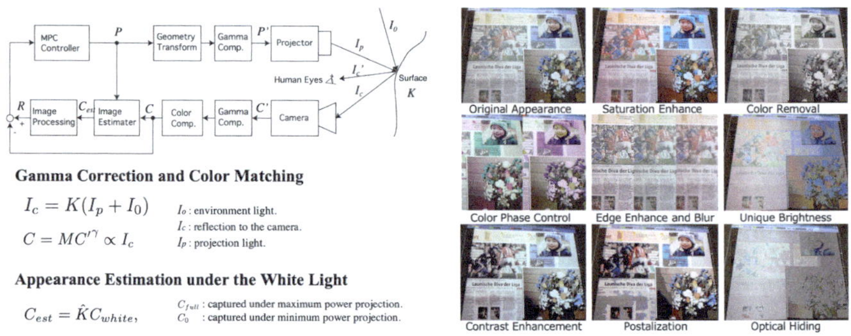

Fig. 1. Appearance control using projector-camera feedback. The illumination projection from the projector manipulates visible color with model predictive control using a reference derived from user-defined image processing of the estimated original appearance in the feedback loop.

2 Real-Time Appearance Manipulation

Using a projector-camera feedback system as illustrated in Fig. 1, we achieved real-time manipulation of visible color and perceptual material appearance [11].

On the object's surface, the reflected light the camera observes I_C and later takes the difference with the control reference given by R and adjusts projection image P along with the Model Predictive Control. In this feedback loop, the $\boldsymbol{R}$ is generated with user-defined image processing by editing the true appearance C_{est}, which is given from the reflectance estimation using the current $\boldsymbol{C}$ and $\boldsymbol{P}$.

This approach, called Appearance Control, enhances visibility for individuals with vision impairments, such as cataracts or color vision deficiencies, and supports night driving with pixelated headlights [12], not simply alternating visible color.

A co-axial aligned projector-camera optical setup eliminates the need for geometric alignment and 3D modeling, enabling successive adaptation to manipulate perceptual material properties, including glossiness and translucency, on arbitrary 3D objects [13].

3 Viewing Direction-Dependent Appearance Manipulation

Conventional projection methods based on Lambertian assumptions cannot account for illumination variations dependent on viewing direction. To overcome this, we developed Viewing Direction-Dependent Appearance Manipulation (VDDAM) techniques that dynamically alter perceived color and illumination based on an observer's viewpoint on non-Lambertian surfaces.

Our initial implementation [14] employed four independently controlled projector–camera feedback units, as shown in Fig. 2. This system achieved structural color-like effects by applying different phase-shifted color parameters to each unit. However, due to the lack of accurate reflectance analysis, it could not reproduce physically faithful BRDFs.

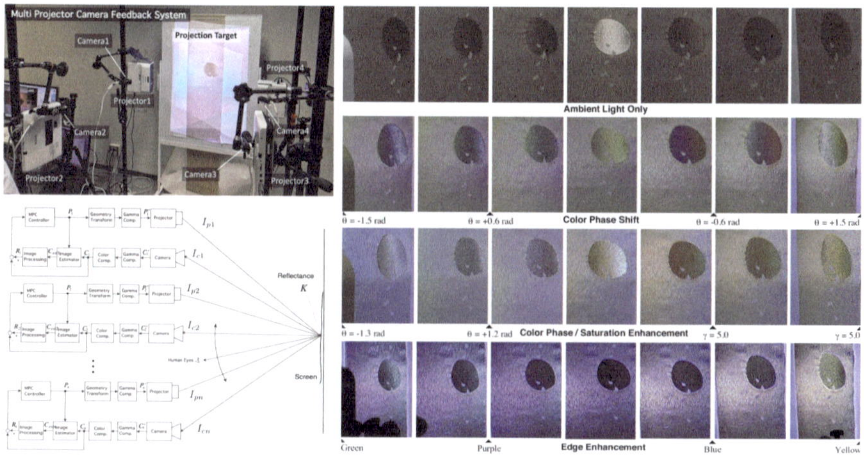

Fig. 2. Direction-dependent appearance manipulation using a four-pair projector-camera feedback system. The drawing foil textile exhibits a beautiful chroma transition along the viewing direction, similar to the structural colors. This effect is achieved through appearance manipulation, combining color phase-shift, saturation enhancement, and edge enhancement.

To overcome these challenges, we introduced a light-field projection method based on directional reflectance analysis [15]. This system utilized a linearly aligned projector array and reflectance matrices to compute projection images through non-negative optimization. The method was later extended to manipulate anisotropic reflection properties by parameterizing the reflectance matrices using the Ashikhmin BRDF model [16].

However, environmental illumination affects its manipulation results. To address this issue, we developed a hybrid method that integrates reflectance analysis with light-field feedback [17, 18]. By employing a multiple-input, multiple-output (MIMO) configuration stabilized through pseudo-inverse decoupling, the system achieved robust manipulation of surfaces with mixed reflectance properties, including mirror-like and retroreflective effects.

These approaches facilitated robust viewing-direction-dependent appearance manipulation on surfaces with mixed reflection properties while compensating for environmental disturbances. However, like earlier methods, this approach required complex geometrical calibration and reflectance analysis, which limited its adaptability.

4 Adaptive BRDF Display using Light Ray Projection

While VDDAM based on the reflectance analysis ensures accurate light distribution to achieve desired BRDFs, it requires re-calibration of geometric and optical parameters when the object's shape or position changes, limiting its flexibility. To overcome this limitation, we proposed an adaptive BRDF display method using mirror-array light-field projection [19].

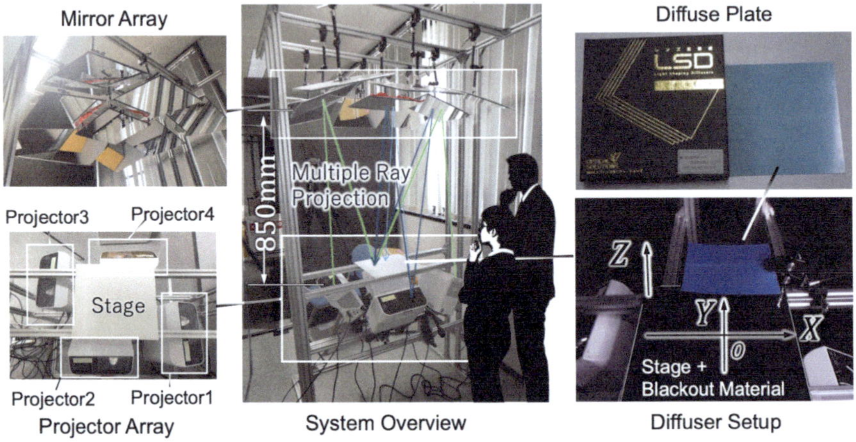

Fig. 3. Our light field projection system consists of four projectors and nine mirrors arranged in a 3 × 3 array. A diffuse plate is placed 150 mm above the stage as shown on the right.

In this approach, we employed the light field projection system [20] consisting of four projectors and nine mirrors arranged in a 3×3 array as illustrated in Fig. 3. Light rays projected from each projector are reflected on the mirrors and overlapped onto the surface, creating dense light ray distributions (3-15 rays per mm2) on the display object.

With this dense light ray casting, the system successfully reproduces the reflected ray distribution described by BRDF, including color phase shifts around specular highlights observed in structural color objects. This approach achieves successive application on the arbitrarily shaped surfaces with flat shape approximation on the generation of the projection light field (Fig. 4). While concerns exist regarding the adequacy of the flat mirror reflection assumption for arbitrary-shaped objects, which could result in color phase gradation inversion or slope changes, determining the appropriate gradation for each BRDF model within the limits of human visual perception remains a challenge. Nevertheless, our experiments confirm that the differences between this approximation and optically correct light-field projection based on surface normal are negligible [21]. This validation ensures that the method effectively adds the desired BRDF materiality to target objects while maintaining visual fidelity.

A key challenge is the sparse angular resolution of the projected light rays, which makes fake spikes in the representation of matte BRDFs. While increasing the number of projectors and mirrors could resolve this, it incurs higher costs. To address this, we employed light-field diffusion using a diffuse plate above the object. Although this method reduces spatial resolution, pre-compensation applied to the BRDF data mitigates the issue effectively [22].

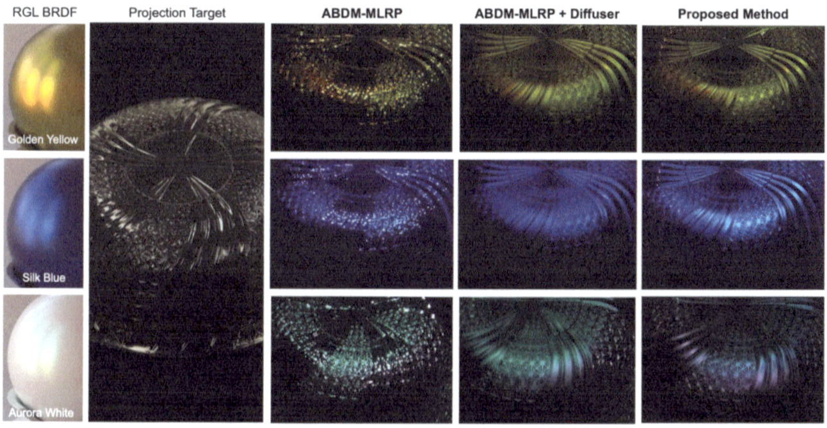

Fig. 4. ABDM-MLRP shows the projection results on the glass bowl with the original method [19]. Adding a diffuser (ABDM-MLRP + Diffuser) removed scattering fake illumination spikes observed in ABDM-MLRP while maintaining the color gradation expressed in BRDF. However, it reduces clarity compared to the proposed method in [22].

5 Conclusion

Our approaches advance the field of Spatial Augmented Reality (SAR) by introducing novel methods for real-time appearance manipulation, viewing direction-dependent appearance manipulation, and adaptive light-field projection. The advancements presented in this paper enable seamless interaction without the need for wearable or handheld devices and have profound implications for a wide range of applications, including product design, optical restoration, immersive museum exhibits, and theater lighting. We believe these techniques allow precise control over perceptual material properties, effectively bridging the gap between the physical and digital worlds and contributing to the realization of the real-world metaverse.

Acknowledgements

This work was supported by JSPS KAKENHI Grant Numbers JP24K02979, JP 25H01886.

References

1. Raskar, R., Welch, G., Low, K., Bandyopadhyay, D.: Shader lamps: Animating real objects with image-based illumination. Proceedings of the 12th Eurographics Workshop on Rendering Techniques, pp. 89–102 (2001).
2. Bimber O., Iwai D., Wetzstein G., Grundhöfer, A.: The visual computing of projector-camera systems. Computer Graphics Forum 27, 8, pp.2219–2245 (2008).

3. Grundhöfer, A., Iwai, D.: Recent Advances in projection mapping algorithms, hardware and applications. Computer Graphics Forum 37, 2, pp.654–675. (2018).
4. Iwai, D.: Projection mapping technologies: A review of current trends and future directions. Proc. Jpn. Acad. Ser. B. Phys. Biol. Sci., vol. 100, no. 3, pp. 234–251 (2024).
5. Okazaki, T., Okatani, T., and Deguchi, K.: A Projector-Camera System for High-Quality Synthesis of Virtual Reflectance on Real Object Surfaces. IPSJ Transactions on Computer Vision and Applications 2, 71–83 (2010).
6. Siegl, C. et al.: Real-time Pixel Luminance Optimization for Dynamic Multiprojection Mapping. ACM Trans. Graph., vol. 34, no. 6, pp. 237:1–237:11 (2015).
7. Nomoto, T., Koishihara, R., Watanabe, Y.: Realistic Dynamic Projection Mapping Using Real-Time Ray Tracing. ACM SIGGRAPH 2020 Emerging Technology SIGGRAPH 2020 (2020).
8. Aliaga, D. G., Law, A. J., and Yeung, Y. H.: A virtual restoration stage for realworld objects. ACM Transactions on Graphics 27, 1 (2008).
9. Shimazu, S., Iwai, D., and Sato, K.: 3D high dynamic range display system. 10th IEEE International Symposium on Mixed and Augmented Reality, 235–236 (2011).
10. Bermano, A., Brüschweiler, P., Grundhöfer, A., Iwai, D., Bickel, B., Gross, M.: Augmenting physical avatars using projector-based illumination. ACM Trans. Graph. 32, 6, 189:1–189:10 (2013).
11. Amano, T., Kato, H.: Appearance Control by Projector-Camera Feedback for Visually Impaired. In: IEEE Computer Society Conference on Computer Vision and Pattern Recognition Workshops PROCAMS 2010, pp.1277–1283 (2010).
12. Amano, T., Takemine, K., Tsukuma, J.: Automotive Lighting Technology for Driver's Visibility Improvement Using Imperceptible Pattern Illumination. In: Proceedings of the International Display Workshops (IDW 2021), pp. 654–657 (2021).
13. Amano, T.: Projection-Based Real-Time Material Appearance Manipulation. In: IEEE Computer Society Conference on Computer Vision and Pattern Recognition Workshops (CCD 2013), pp. 918–923 (2013).
14. Amano, T., Ushida, S., Miyabayashi, Y.: Viewpoint-Dependent Appearance Manipulation with Multiple Projector-Camera Systems. In: International Conference on Artificial Reality and Telexistence and Eurographics Symposium on Virtual Environments (ICAT-EGVE 2017), pp. 101–107 (2017).
15. Amano, T., Minami, K.: Structural Color Display on Retro-Reflective Objects. In: 25th International Conference on Artificial Reality and Telexistence and 20th Eurographics Symposium on Virtual Environments (ICAT-EGVE 2015), pp. 37–44 (2015).
16. Ohsumi, S., Amano, T.: Manipulation of Anisotropic Reflections Based on Optical Models Using Multiple Projectors. In: CEUR Workshop Proceedings, vol. 3297, pp. 3–8 (2022).
17. Amano, T., Yoshioka, H.: Viewing-Direction Dependent Appearance Manipulation Based on Light-Field Feedback. In: Lecture Notes in Computer Science (LNCS), vol. 12499, pp. 192–205 (2020).
18. Matsumoto, Y., Amano, T.: A Study on Shadow Removal for Viewing Direction-Dependent Appearance Manipulation. In: 18th Annual EuroXR International Conference (EuroXR 2021), virtual, pp. 21–26 (2021).
19. Kimura, K., Amano, T.: Perceptual BRDF Manipulation by 4-Degree of Freedom Light Field Projection Using Multiple Mirrors and Projectors. In: 19th Annual EuroXR International Conference (EuroXR 2022), pp. 95–99 (2022).
20. Raich, K., Amano, T.: Reproduction of Multiple Mirror-Based Arbitrary Lighting Environment. In: 19th Annual EuroXR International Conference (EuroXR 2022), pp. 81–86 (2022).

21. Kodan, K., Amano, T.: Texture reproducibility evaluation on BRDF reproduction by Light Field Projection. In: Proceedings of the 16th Asia-Pacific Workshop on Mixed and Augmented Reality (APMAR 2024) (2024).
22. Amano, T., Nishida, S.: Enhancing Materiality in Adaptive BRDF Display with Light Ray Diffusion. In: Virtual Reality and Mixed Reality: 21st EuroXR International Conference (EuroXR 2024), Proceedings, pp. 134–151. Springer-Verlag, Berlin, Heidelberg (2024).

Contact Lens Display Using Holographic Image Generation

Yasuhiro Takaki

[1] Tokyo University of Agriculture and Technology, 2-24-16, Nakacho, Koganei, Tokyo, Japan
ytakaki@cc.tuat.ac.jp

Abstract. Recently, the smart contact lenses attract a lot of interests, especially, the contact lens display which can be used for an AR visual interface device. Recently, the contact lens display using the holographic image generation technique has been proposed. In this paper, the developments of the holographic contact lens display are explained. The comparisons of the image generation techniques for the contact lens display are also provided. The R&D project of the holographic contact lens display in Japan is explained.

Keywords: Contact Lens Display, Holography, AR.

1 Introduction

Recently, the developments of new-generation mobile devices have been conducted actively. Among then, the contact lens display is the final form of visual interface devices which are non-invasive to human. Because eyes cannot focus on the display screen embedded in a contact lens, the image generation technique specialized for the contact lens displays is needed and several techniques have been proposed [1–4]. This article explains the use of the holographic technique for the image generation [1]. The comparisons of the holographic technique [1] and the other techniques [2–6] are provided. Moreover, the R&D project for the holographic contact lens display supported by Japanese government is explained.

2 Holographic Contact Lens Display

Images should be generated outside the screen of the display device embedded in a contact lens so that eyes can focus on the images. The use of the holographic technique can address this issue considering that a three-dimensional (3D) image can be generated apart from a hologram. As shown in Fig. 1, a hologram pattern is displayed on a display embedded in a contact lens. The laser light illumination on the hologram pattern generates a wavefront, which produces a 3D image at a distance at which the eye can focus. In addition, a phase-only hologram is used, which modules a phase distribution of light. Because the phase-only hologram does not modulate an amplitude distribution, it is a transparent hologram. Thus, the see-through function is provided.

S. Kurimura and T. Yagi (eds.), *Proceedings of Laser Display and Lighting Conference 2025*, Springer Proceedings in Physics 429,
https://doi.org/10.1007/978-981-96-9782-3_32

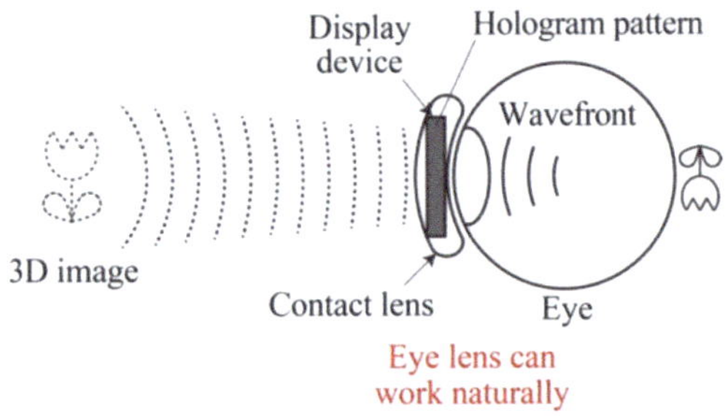

Fig. 1. Holographic contact lens display.

Figure 2 shows the structure of the holographic contact lens display [1]. The phase-only spatial light modulator (SLM) modulates the horizontally polarized light, and the laser backlight emits a laser light that is polarized in this direction to illuminate the phase-only SLM. The polarizer transmits a vertically polarized light to prevent an external light, from the outer scene, to be modulated by the phase-only SLM. Thus, the holographic image produced by the SLM and the light from the outer scene have orthogonally polarized directions and can be simultaneously viewed, providing the see-through function. In this explanation, the horizontal and vertical polarization directions can be interchanged.

The experimental verification of the holographic technique was shown in Ref. 1.

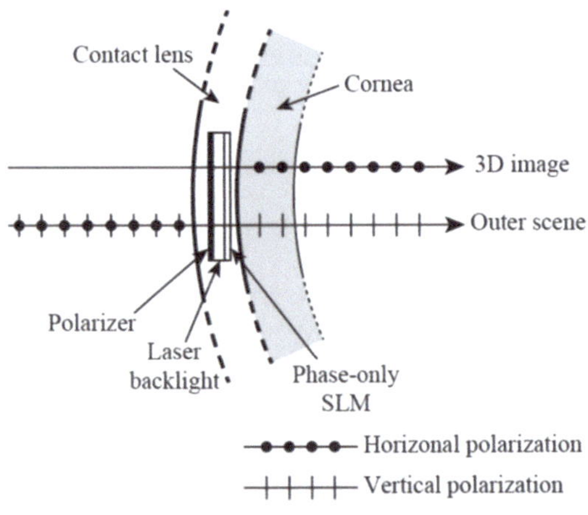

Fig. 2. Structure of holographic contact lens display.

3 Comparisons of Image Generation Techniques

Beside the holographic technique [1], for the image generation techniques used for the contact lens displays, the pixel-by-pixel technique [2–4] and the tiny-projector technique [5, 6] have been developed.

The pixel-by-pixel technique attaches a micro-optical element to each pixel of the display device to converge light on the retina. The use of a Fresnel lens and an optical fiber as the micro-optical element was proposed in Refs. 2 and 3, respectively. In Ref.4, the use of a meta-surface was also proposed and the see-through function was experimentally verified.

The prototype system of the contact lens display using the tiny-projector technique has been demonstrated [5, 6]. The tiny projector consisted of a micro-LED display and a projection optics based on Cassegrain telescope. Because the projector had a certain depth, a scleral lens, which is a large rigid-surface contact lens, was used.

To enable the widespread use of holographic contact lens displays, we think that they should be realized as soft contact lenses. Considering that the thickness of the center part of the soft contact lenses (optic zone) is 0.15 ~ 0.25 mm and the outer radius of the cornea is 7.8 mm, the dimensions of the SLM, which can be placed inside the soft contact lens, are limited. The possible combinations of the pixel pitch, resolution, and screen size of the display devices for a thickness of 0.15 mm are listed in Table 1.

Table 1. Parameters of display that can be embedded in a contact lens of thickness of 0.15 mm.

Pixel pitch [μm]	Resolution	Screen size [mm^2]
3	256×256	0.768×0.768
	512×512	1.54×1.54
	1,024×1,024	3.07×3.07
5	256×256	1.28×1.28
	512×512	2.56×2.56
8	256×256	2.05×2.05

The comparisons of the three image generation techniques for the contact lens displays are shown in Table 2. For the pixel-by-pixel technique, because the micro-optical elements can be attached to the pixels of the display, the structure can be thick. For the tiny-projector technique, because the distance between the projection lens and the display should be larger than the focal length of the projection lens, the reduction of the system depth is limited. For the holographic technique, because the laser backlight can be tightly combined with the phase-only SLM, the system depth can be thin. Although the holographic technique can electronically control the image distance, the laser backlight is required and the hologram calculation requires a large computation load. For the pixel-by-pixel technique, the micro-optical elements can be directly attached to the pixels only when the pixels emit plane waves. For the pixel-by-pixel and the tiny projector techniques, since the display is opaque, the display screen size cannot be increased regarding to the loss of the light from the outer scene. For the holographic technique, the screen size can be increased because the phase-only SLM is transparent.

The resolution of the retinal images generated by the three image generation techniques is considered. The resolvable length of retinal images depends on the extent of light distribution on a cornea. For the pixel-by-pixel technique, the width of the micro-optical element, which is up to the pixel pitch, determines the resolvable length. When the pixel pitch is denoted by p, the wavelength of light is denoted by λ, and the focal length of eyes is denoted by f, the resolvable length is given by $\lambda f / p$. For the tiny-projector technique, the diameter of the projection lens determines the resolvable length as $\lambda f / d$ when the diameter of the projector lens is denoted by d. For the holographic technique, because a whole hologram functions as a single optical element, the width of the hologram determines the resolvable length, which is given by $\lambda f / w$ when the width of the hologram is denoted by w. Therefore, the pixel-by-pixel technique cannot produce high-resolution retinal images even when the pixel pitch is reduced. The tiny-projector and holographic techniques can produce high-resolution retinal images. When the projection lens diameter and the hologram width is 2 mm, the resolvable length becomes 4 μm (λ = 0.5 μm and f = 16 mm), which is comparable to the resolvable length of the retina. When the pixel pitch is p = 5 μm, the resolvable length becomes 1.6 mm for the pixel-by-pixel technique.

Table 2. Comparisons of image generation techniques used for contact lens displays.

	Pixel-by-pixel	Tiny projector	Holography
Structure	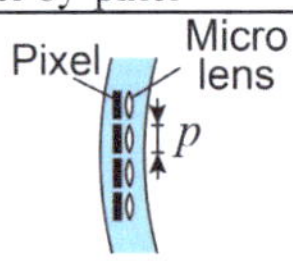	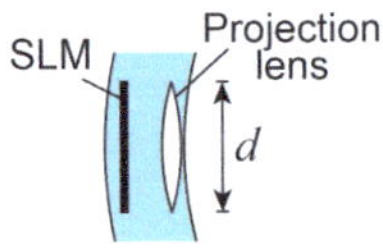	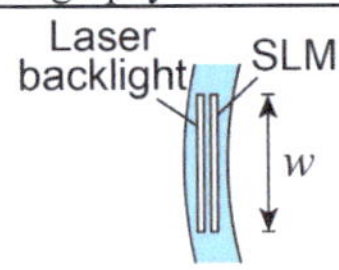
Thickness	Thin < 1mm	Thick > 1 mm	Thin < 1 mm
Resolution	Low	High	High
Image distance	Pan-focus	Fixed	Variable
Backlight	Not required	Not required	Required
Calculation	None	None	Heavy

4 R&D Project for Holographic Contact Lens Display

The holographic contact lens display provides both the thick structure and the high-resolution. A new research project named "Developments of Innovative and Fundamental Technologies for Holographic Contact Lens Display" has started on April 2024, which is funded by National Institute of Information and Communications Technology (NICT), Japan. As shown in Fig. 3, five organizations from academia and industry participate in this project. The project consists of five R&D targets, such as, the image formation technique including laser backlight and hologram calculation technique (target 1), the ultra-compact and ultra-thin spatial light modulator (target 2), the ultra-compact electronic devices including antenna and rotation angle sensor (target 3), the integration to contact lens including structure development and waterproof technique (target 4), and the evaluation of effects on visual system (target 5).

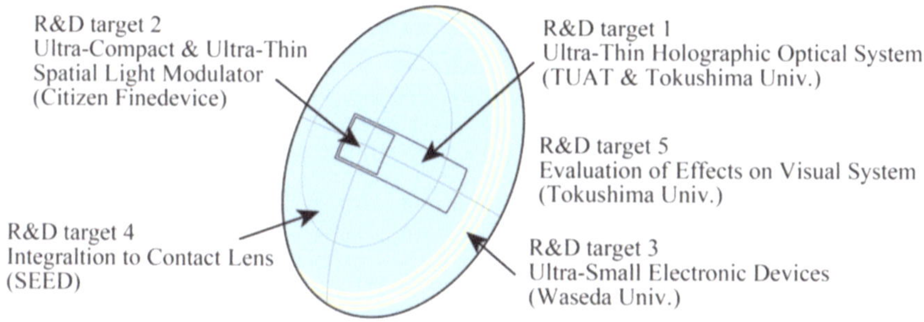

Fig. 3. R&D project for holographic contact lens display.

5 Conclusions

The holographic implementation of the contact lens display was explained, and the comparisons to the other image generation techniques for the contact lens displays were shown. The holographic technique has the possibility to provide high-resolution retinal images with an ultra-thin form factor.

Parts of these research results were obtained from the commissioned research (JPJ012368C08601) by National Institute of Information and Communications Technology (NICT), Japan.

References

1. J. Sano and Y. Takaki, "Holographic contact lens display that provides focusable images for eyes," Optics Express, **29**,10568-10579 (2021).
2. A. R. Lingley, M. Ali, Y. Liao, R. Mirjalili, M. Klonner, M. Sopanen, S. Suihkonen, T. Shen, B. P. Otis, H. Lipsanen, and B. A. Parviz: "A Single-Pixel Wireless Contact Lens Display," J. Micromech. Microeng., **21**, 125014 (2011).
3. C. Chen, L. Mi, C. P. Chen, H. Liu, J. Jiang, and W. Zhang, "Design of foveated contact lens display for augmented reality," Opt. Express **27**, 38204-38219 (2019).
4. S. Lan, X. Zhang, M. Taghinejad, S. Rodrigues, K. Lee, Z. Liu and W. Cai, "Metasurfaces for Near-Eye Augmented Reality," ACS Photonics, **6**, 864-870 (2019).
5. M. Wiemer, R. Winoto, "Mojo Lens - AR contact lenses for real people," 2021 IEEE Hot Chips 33 Symposium (HCS), Palo Alto, CA, USA, 1-56 (2021).
6. P. S. Martin, "Contact lens embedded microLED displays," SID Symposium Digest of Technical Papers **53**, 1521-1523 (2022).

Part VI
Novel and Emerging Technologies LDC6

Temperature distribution of AMOLED panel with the consideration of IR-drop

Qibin Feng[1-2], Baoceng Wang[1-2], Zi Wang[1-2], Guoqiang Lyue[1-2]

[1] Hefei University of Technology, Tunxi Rd 193, 230009 Hefei, China
[2] Anhui Province Key Laboratory of Measuring Theory and Precision Instrument, School of Instrument Science and Optoelectronics Engineering, Hefei University of Technology, Tunxi Rd 193, 230009 Hefei, China
fengqibin@hfut.edu.cn

Abstract. Active-matrix organic light-emitting diodes (AMOLEDs) present big challenge to liquid crystal displays (LCDs). However the illumination efficiency of an AMOLED heavily depends on temperature. The accurate temperature distribution is important for implementing compensation of AMOLED panel to improve display uniformity. With the increase of size and resolution, the voltage drop (IR-drop) caused by the resistance of the power supply line can't be ignored, which also has influence on temperature distribution. The paper proposes a temperature distribution analysis method based on IR-drop. Firstly, sparse-representation-based IR-drop model is put forward, which can help solve IR-drop of AMOLED with high resolution. Secondly, the paper introduces IR-drop based heat power model, which can help perform the temperature simulation. Finally, the temperature distribution of the AMOLED panel is obtained by finite element analysis. The temperature measurements are performed on a 27inch 4K AMOLED panel to validate the proposed method. The simulation results are in good agreement with the actual measurements, which means the analysis method proposed in this paper presents high accuracy and high practicability.

Keywords: AMOLED, IR-drop, Temperature distribution.

1 Introduction

With many merits such as high-dynamic range, fast response time, and ultra-thinness, active-matrix organic light-emitting diodes (AMOLEDs) present big challengence to liquid crystal displays (LCDs) [1,2]. Although the brightness and luminous efficiency of AMOLEDs have achieved great improvement, the lifetime and stability are still important bottlenecks limiting its applications [3].

As the fact that the luminous efficiency of an AMOLED heavily depends on temperature, the nonuniformity of temperature distribution would cause the nonuniformity of illuminance [4]. In order to improve the thermal stability of AMOLED panels, there exist two kinds of methods: one is to develop high-temperature-resistant organic materials, such as hole-transporting materials with high glass transition temperatures [5] and another is to establish efficient heat dissipation

S. Kurimura and T. Yagi (eds.), *Proceedings of Laser Display and Lighting Conference 2025*, Springer Proceedings in Physics 429,
https://doi.org/10.1007/978-981-96-9782-3_33

pathways [6]. For the second method, the precise temperature distribution of an AMOLED panel is precondition to implement efficient compensation. Nowadays most researchers employ the finite element analysis method to get the temperature distribution of an AMOLED panel [7-8]. However, there is almost no consideration of the effect of IR-drop on the temperature distribution.

IR-drop is the voltage drop along the power metal line from the driver edge to the far end edge of the AMOLED panel, which leads to emission nonuniformity [9-10]. When the size and resolution of AMOLEDs develop toward larger and higher, IR-drop becomes unignorable. The paper proposes the IR-drop based heat power model, which can help achieve the accurate temperature distribution of a medium- or large-size and high-resolution AMOLED panel.

2 IR-drop-based heat power model of AMOLED

2.1 Sparse-representation-based IR-drop model

With the consideration that the capacitance and inductance are usually ignored when performing static IR-drop analysis, the equivalent circuit model of the driving circuit of an AMOLED panel can be regarded as a purely resistive network, as shown in Fig. 1. R_{MD} and R_{TD} are the thin layer resistances of the power metal line segments between two adjacent sub-pixels along the latitude and longitude, which can be obtained according to the following equations:

$$R_{TD} = \rho \times \frac{l_{MD}}{h_{TD} \times w_{TD}} \tag{1}$$

$$R_{MD} = \rho \times \frac{l_{TD}}{h_{MD} \times w_{MD}} \tag{2}$$

where l_{TD} and l_{MD} denote the lengths of the latitudinal and longitudinal power metal lines, respectively. h_{TD} and h_{MD} denote the line heights of the latitudinal and longitudinal power metal lines, respectively. w_{TD} and w_{MD} denote the line widths of the latitudinal and longitudinal power metal lines, respectively. ρ denotes the resistivity of the power metal lines.

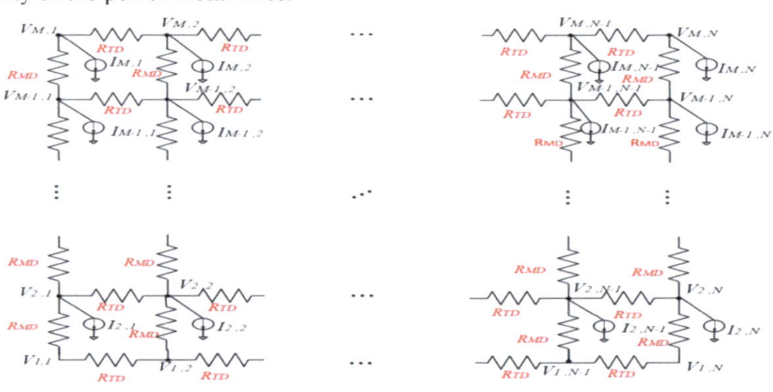

Fig. 1. The equivalent circuit model of the driver circuit for AMOLED panel.

According to the equivalent circuit theory, the equivalent circuit model of the driving circuit of the AMOLED panel can be established as an IR-drop model represented by the vector matrix equation, as shown in (3).

$$\begin{bmatrix} G'_{1,1} & C_1 & 0 & \cdots & 0 \\ C_1 & G'_{2,2} & C_2 & \ddots & \vdots \\ 0 & C_2 & \ddots & \ddots & 0 \\ \vdots & \ddots & \ddots & G'_{M-1,\ M-1} & C_{M-1} \\ 0 & \cdots & 0 & C_{M-1} & G'_{M,M} \end{bmatrix} \begin{bmatrix} V_1 \\ V_2 \\ \vdots \\ V_{M-1} \\ V_M \end{bmatrix} = \begin{bmatrix} I_1 \\ I_2 \\ \vdots \\ I_{M-1} \\ I_M \end{bmatrix} \tag{3}$$

where G is an (M×N)×(M×N) matrix that represents the conductance matrix for the interconnected resistors, V is an (M×N)×1 matrix that represents the vector matrix of all AMOLED node voltages, and I is an (M×N)×1 matrix that represents the vector matrix of all AMOLED node currents. C is an N×N matrix.

The sub-matrix $G'_{i,i}$ represents the connectivity in the same row, and the value on the diagonal is the self-conductance of each node (i, j) in row i. The sub-diagonal value is the mutual conductance, which is negative. The sub-matrix C_i represents the connectivity of two neighboring rows, the non-zero elements of the sub-matrix C_i are all located on its diagonal, and its value is the conductance of the two neighboring rows, which is negative.

Take $G'_{1,1}$ as an example. As shown in the equation (4), all elements are zero except for those on the diagonal. With the increase of resolution and size, the IR-drop model of an AMOLED has a high matrix order, which requires larger running memory and longer running time. By exploiting the sparse representation in the field of artificial intelligence, only the non-zero elements in the conductivity matrix G are stored, which can help reduce the size of the matrix and make it possible to accurately solve the IR-drop of an AMOLED panel with large size and high-resolution. With the introduction of sparse representation, the first column can be expressed as the equation (5).

$$G'_{1,1} = \begin{bmatrix} 1 & 0 & 0 \\ 0 & \ddots & 0 \\ 0 & 0 & 1 \end{bmatrix} \tag{4}$$

$$G_1 = \begin{bmatrix} G'_{1,1} \\ C_1 \\ 0 \\ \vdots \\ 0 \end{bmatrix} \xrightarrow{\text{Sparse representation}} \begin{bmatrix} (1,1)G''_{1,1} \\ (2,1)g_{TD} \\ (n+1,1)g_{MD} \\ \vdots \\ (n-1,n)g_{TD} \\ (n,n)G''_{1,n} \\ (2n,n)g_{MD} \end{bmatrix} \tag{5}$$

In order to solve V and I in the IR-drop model, the exact $I - V$ relationship has to be firstly determined. I–V relationship of driving TFT commonly used in AMOLED is expressed as the following [11-12]:

$$I_{i,j} = I_{SD} = \frac{1}{2}\mu_n C_{ox}\frac{W}{L}(V_{SG} - |V_{th-DTFT}|)^2(1 + \lambda V_{SD}) \tag{6}$$

where μ_n, C_{ox}, $\frac{W}{L}$ are the mobility, oxide capacitance, and aspect ratio of the driving TFT, respectively. $\mu_n C_{ox}\frac{W}{L}$ is set to $200\frac{\mu A}{V^2}$. $V_{th-DTFT}$ is V_{th} of the driving TFT, and λ is the channel length modulation coefficient. V_{SG}, V_{SD} are the source-to-gate voltage, and source-to-drain voltage, respectively.

$I_{i,\ j}$ and $V_{i,\ j}$ can be got by interactively solving (3). The conductance matrix G of (3) is a positive definite matrix. If the iterations converge, the solution for V^{k+1} should be closer to the solution of the original equation than for V^k, so a more accurate value of the node voltages can be obtained by replacing V^k by V^{k+1} .The continuous iterations are designed herein to stop as the difference between two consecutives V's through the iterations becomes smaller than the iteration error. The iteration result can be used as the real node voltage value for IR-drop analysis, as shown in following.

$$\delta_{i,j} = V_{dd} - V_{i,j} \tag{7}$$

where $\delta_{i,j}$ is the IR-drop from V_{dd} to the node (i, j).

2.2 IR-drop based heat power model

After obtaining the IR-drop at each node of the AMOLED panel, IR-drop based heat power model is put forward. The output of the power model is used as the input to the AMOLED thermal simulation in order to analyze the temperature distribution. The energy conversion efficiency is set to $\eta\%$, which means that $(1-\eta)\ \%$ of the electrical power is converted to heat. The IR-drop based heat power model can be expressed by (8)- (10).

$$P_{OLED(i,j)} = V_{OLED(i,j)} * I_{OLED(i,j)} * \ (1-\eta)\ \% \tag{8}$$

$$V_{OLED(i,j)} = V_{i,j} - V_{SS} - V_{SD} \tag{9}$$

$$I_{OLED\ (i,j)} = I_{i,j} \tag{10}$$

where $P_{OLED\ (i,j)}$ is the thermal input power of the node $(i,\ j)$. $V_{OLED\ (i,j)}$ is the voltage of the node $(i,\ j)$. $I_{OLED\ (i,\ j)}$ is the current flowing through the node $(i,\ j)$.

3 Simulations

3.1 IR-drop solution

IR-drop on each pixel is firstly solved for a 27inch 4K (2160×4096) AMOLED panel. V_{dd} on the driver side of the panel is set to 24 V. R_{TD} and R_{MD} are set to 0.5 Ω. The distribution of $V_{i,\ j}$ for the AMOLED panel at the grayscale of 255 is shown in Fig. 2.

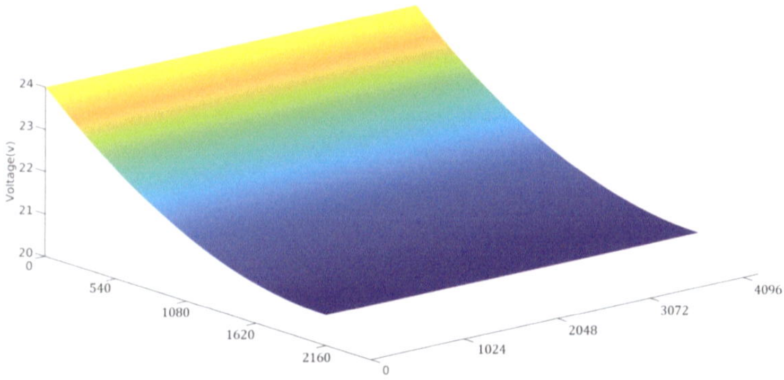

Fig. 2. Results of $V_{i,\ j}$ distribution in OLED panel with 4K resolution.

3.2 Temperature distribution simulation

The panel model and parameters of an AMOLED are shown in Fig. 3 and Table 1. The panel model consists of glass substrate (0.7 mm), OCA1 layer (0.1 mm), POL layer (0.02 mm), OCA2 layer (0.1 mm), OLED layer (0.1 mm), metal cover (0.15mm), thermal grease layer (0.08 mm) and the back frame (2.5 mm). The OLED layer consists of organic light-emitting layer, the thin-film circuit layer, and the silicon oxide insulation layer.

When displaying an all-white image (255 grayscale at each pixel), the simulated temperatures of 8 points on AMOLED from the driver edge to far end edge are taken out and shown in Table 2. For better comparison, the temperatures without the consideration of IR-drop are also presented. It can be seen when considering no IR-drop, the temperature on each point is same. When considering IR-drop, the temperature gradually deceases from P1 to P8, which means the temperature drops along the direction from driver edge to far end edge. Such downward trend coincides exactly with the trend of the IR drop, which well demonstrates the impact of IR-drop on the temperature distribution.

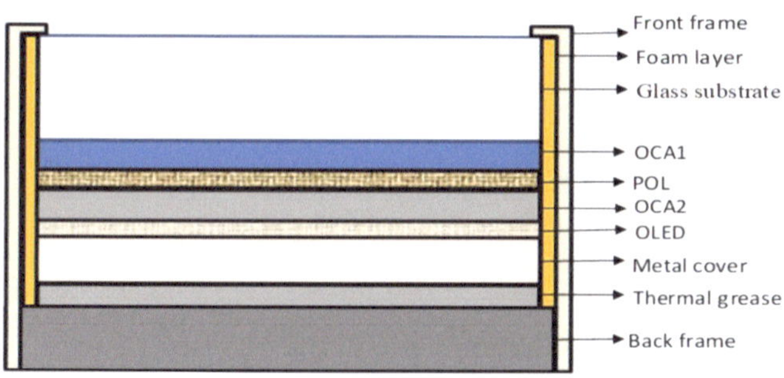

Fig. 3. Structural diagram of an AMOLED panel.

Table 1. Simulation parameters.

Structure	Material	Thermal conductivity $(W/m{\cdot}K)$
Glass substrate	Glass	1.1
OCA1	Olefin film	0.24
POL	Polymer film	0.2
OCA2	Olefin film	0.24
OLED	Nichrome film	43.8
Metal cover	Stainless steel	26.3
Thermal grease	Aluminum	1
Back frame	ABS	0.24
Front frame	ABS	0.24
Foam	EPE	0.04

Table 2. Simulated temperatures of 8 points on AMOLED (°C)

Points	Temperature with IR-drop	Temperature without IR-drop	Difference
P1	32.4	32.4	0
P2	31.7	32.4	0.7
P3	30.9	32.4	1.5
P4	30.1	32.4	2.3
P5	29.3	32.4	3.1
P6	28.6	32.4	3.8
P7	27.9	32.4	4.5
P8	27.0	32.4	5.4

4 Experimental validation

An 27inch AMOLED with the resolution of 4K was employed to validate the proposed method. As the Fig. 4 shown, 8 resistance strain gauges were firmly attached on the upper surface of AMOLED. The measure temperatures of 8 points are presented in Table 3.

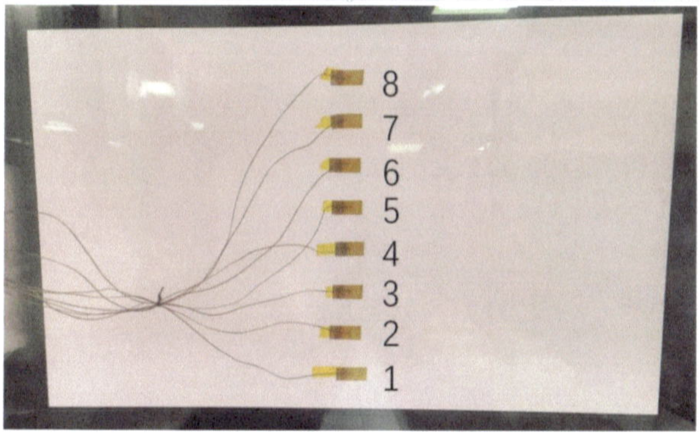

Fig. 4. Temperature measurement of AMOLED panel.

Table 3. Measured temperatures of 8 points on AMOLED (°C)

P1	P2	P3	P4	P5	P6	P7	P8
33.2	32.5	31.7	31.2	30.6	29.8	28.8	27.4

The data in Table 3 shows that the temperature difference between P1 and P8 is 5.8 °C, which clearly shows the presence of IR-drop. The simulated and measured temperatures are both illuminated in Fig. 5.

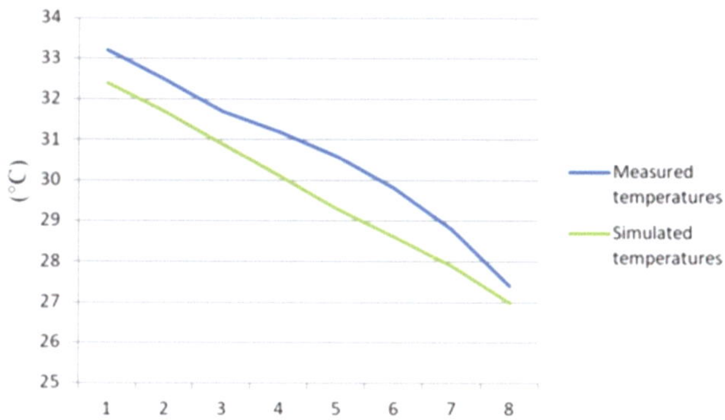

Fig. 5. Comparison between measured temperatures and simulated temperatures.

It can be seen that the measured temperatures are slightly higher than the simulated temperatures, which may be caused by the difference between the real and simulated convection condition. The largest difference of 1.3 °C between the measured and simulated temperatures happens at the point 5, which is 4.25% error and presents high accuracy of the proposed method.

5 Conclusions

This paper proposes an AMOLED temperature distribution analysis method based on IR-drop, which can effectively achieve the temperature distribution for medium- or large-size and high-resolution AMOLED panels. The good agreement between the simulated and measured temperatures demonstrates the analysis method proposed in this paper high accuracy and strong practicality.

Funding: This research was funded by Major Science and Technology Research Project of Anhui Province, grant number 2023z020002.

References

1. Cheng, Y: Process in display technologies and their application, Chinese Optics 3, 31–36 (2010).
2. Guan, Y.X.; Chen, L.J.; Chen, P.: Organic light-emitting device based on LSMO, Infrared and Laser Engineering, 44, 2105-2109 (2015).
3. Zhang, J.P.; Wen, S.S.: Simulation of thermal properties for polymer light-emitting diodes, Acta Polymerica Sinica, 1458-1463 (2010).
4. Gärditz, C.; Winnacker, A.: Impact of Joule heating on the brightness homogeneity of organic light emitting devices, Applied physics letters, 90,103506 (2010).
5. Xiang, N.; Gao, Z.: Novel fluorene/indole-based hole transport materials with high thermal stability for efficient OLEDs. *Dye. Pigment.*, *137*, 36–42 (2017).
6. Boroumand, F.A.; Hammiche, A.: Characterizing joule heating in polymer light-emitting diodes using a scanning thermal microscope, Advanced Materials, 16, 252-256 (2004).
7. Zhang, W.W.; Wu, Z.X.; Liu, Y.W.: Thermal analysis of organic light emitting diodes based on basic heat transfer theory, Chinese Physics Letters, 32, 143–147 (2015).
8. Li, Y.T.; Song, X.F.; Chen, J.L.: Encapsulation's thermal characteristics for organic electroluminescent devices, Acta Photonica Sinica, 40, 1630-1635 (2010).
9. Yum, J.; Yu, Y.-H.; Park, H.; Lim, H. A Novel Method to Reduce Luminance Variation Due to IR-drop in Active Matrix OLED Displays. In Proceedings of the 2020 IEEE International Conference on Consumer Electronics (ICCE), pp. 1–4 , NV, Las Vegas.
10. Kohno, T.; Kageyama, H.; Miyamoto, M.; Ishii, M.; Kasai, N.; Nakamura, N.; Akimoto, H.: High-Speed Programming Architecture and Image-Sticking Cancellation Technology for High-Resolution Low-Voltage AMOLEDs, IEEE Transactions on Electron Devices, 58, 3444-3452 (2011).
11. Chao, P.C.-P.; Cheng, S.-S.; Chen, C.-H.: A New IR-Drop Model That Improves Effectively the Brightness Uniformity of an AMOLED Panel. *IEEE J. Electron Devices Soc. 10*, 627–636 (2022).
12. Lim, S.-G.; Lee, K.; Kim, Y.-J. Mobile AMOLED Display Power Model Considering I–R Drop in Smartphones. *IEEE Trans. Ind.Electron. 68*, 2694–2702 (2021).

Multi-Gbps Visible Light Communication with High-Efficiency, Low-speckle Contrast White Laser Light

Joseph P. Lazzaro[1[0009-0004-8911-9928]], Xiaoxin Wang[1[0000-0002-8307-6507]], Charles J. Carver[2[0000-0002-6664-1893]], Nicholas Shade[1[0000-0002-9510-5627]], Hadleigh Schwartz[2[0009-0001-6151-2240]], Yanchen Liu[2[0000-0001-6024-5104]], Xia Zhou[2[0000-0002-2852-9024]], Jifeng Liu[1[0000-0003-4379-2928]], Eric R. Fossum[1[0000-0002-6232-0515]]

[1] Dartmouth College, Hanover NH 03755, USA
[2] Columbia University, New York NY 10027, USA

Abstract. Multi-Gbps visible light communication is achieved using direct red-green-blue laser mixing in a novel packaging platform to produce white light with a low-speckle contrast of ≤ 5% and a high luminous efficacy of 150 lm/W_e.

Keywords: Visible Light Communication, Laser Diode Lighting, Multiple Scattering, Speckle Contrast, Direct Color-mixing.

1 Introduction

Global electricity consumption for lighting continues to increase, and this growth is not offset by the improvements in the efficacy of new lighting systems, which remain 30% below the requirements outlined in the International Energy Agency's 2030 Net-Zero Scenario [1]. Efficient white lighting from direct red-green-blue laser mixing has the potential to curb this offset. At the same time, wireless communication needs are increasing, with more devices connected to traditional routers than ever before, further congesting the already crowded radio-frequency spectrum while requiring faster download speeds [2]. Visible light communication (VLC), a technology that uses light to transmit data, is emerging as a solution for more rapid data rates without disturbing or being disturbed by other connected devices. To address both challenges, a multi-Gbps VLC system was developed which utilizes a novel transmitter with low-speckle contrast and high luminous efficacy, enabled by direct color-mixed white light.

2 Lighting System

To achieve high-efficiency, low-speckle contrast dispersed white light, a color-mixed laser diode (CM-LD) prototype was designed and characterized, consisting of RGB LDs, an acrylic mixing rod, two types of diffusers (rod diffuser and output diffuser), and a reflective cavity. Here, RGB laser beams are firstly coupled and guided in the mixing rod. The bulk scattering from the rod-bottom diffusers then redirects the light to the highly reflective interior surfaces of the cavity. The output diffuser allows further

S. Kurimura and T. Yagi (eds.), *Proceedings of Laser Display and Lighting Conference 2025*, Springer Proceedings in Physics 429,
https://doi.org/10.1007/978-981-96-9782-3_34

bulk scattering and forward transmission of the CM-LD white light [3]. With this compact, universal-packaging platform for direct color-mixing schemes, a low-speckle contrast of ≤ 5% and a high luminous efficacy of 150 lm/W_e were achieved. With future improvement of blue-laser-diode efficiency and development of yellow/amber/orange laser diodes, this universal white-laser package can potentially achieve a luminous efficacy > 275 lm/W_e in reference to the input electrical power, ~1.5x higher than state-of-the-art LED lighting and exceeding the target of 249 lm/W_e for 2035 [4]. In addition to its high efficacy, this solution demonstrated a 3 dB bandwidth of 1.3 GHz, indicating great potential for this laser-based white-lighting package in VLC.

3 Communication System

To demonstrate the communication capabilities of the color-mixed white-laser light, a VLC system was developed featuring the mixing package, an APD210-Si avalanche photodiode (APD), two ARM + FPGA development platforms, and two custom printed-circuit boards (PCBs), functioning as the transmitter and receiver, respectively. The first FPGA receives data via direct memory access (DMA) from a desktop computer, encodes it, and outputs it at a programmable data rate of up to 1.5 Gbps to the high-speed transceiver pins on the FPGA. The encoding scheme used here is 8b/10b, and as such, a line rate of 1.25 Gbps is required from the transmitter to achieve a data rate of 1 Gbps due to the two non-data bits in the encoding. The transmitter PCB receives this signal and uses it to modulate the red-laser-diode in the mixing package with on-off keying. The APD on the receiving end detects this optical signal and inputs it to the receiver PCB. The receiver PCB then converts the received signal into one decodable by the second FPGA. The second FPGA moves the decoded message to the processor on the same chip, where it is formatted into packets and delivered via an ethernet cable to a second computer. With this setup, the ability of the system to stream a live webcam video from one computer to another was demonstrated using a data rate of 1 Gbps and a distance of 1 meter between the lighting package and APD (Fig. 1 (a)).

When characterizing the system, AMD's LogiCORE™ IP Integrated Bit Error Ratio Tester (IBERT), which provides pattern generators and checkers implemented in FPGA logic, was employed [5]. The impact of data rate, distance, laser bias current, and angle of incidence on the bit error rate (BER) of the VLC system was investigated. In each test, over 40 billion bits generated from a 7-bit, pseudo-random binary sequence (PRBS-7) on the FPGA were transmitted and received. BERs below the forward-error-correction (FEC) threshold of 3.8×10^{-3} [6] were achieved for data rates up to 1.25 Gbps, free-space link distances up to 3.81 meters, and angles of incidence up to 30 degrees. With a distance of 1 meter between the lighting package and the APD, a BER of 2.4×10^{-11} was measured at a data rate of 1 Gbps, enabling high-performance streaming from a typical desk lamp height to a computer below. Fig. 1 (b) and (c) show a portion of the data collected in this characterization effort.

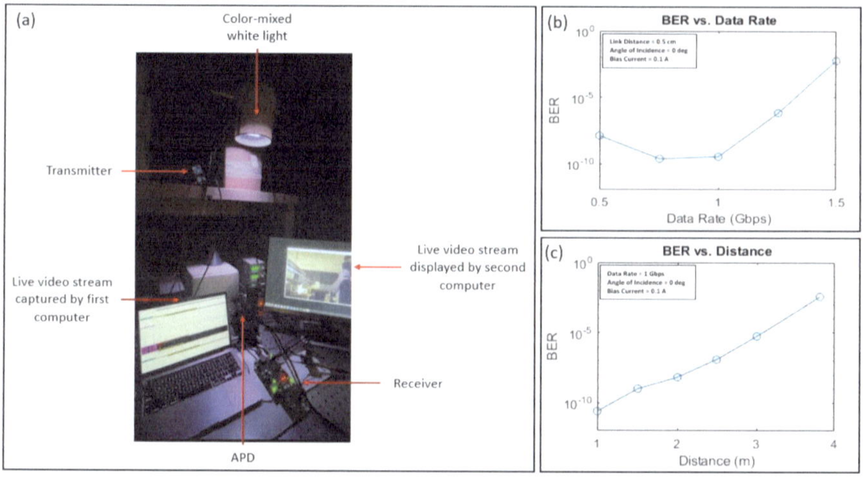

Fig. 1. (a) Live webcam video stream sent over free space by the color-mixed LD transmitter and displayed on a second, receiving computer. The impact of (b) data rate and (c) distance on bit error rate was determined when characterizing the communication system.

4 Conclusion

Using off-the-shelf electronics to modulate a luminaire consisting of RGB LDs, an acrylic mixing rod, two types of diffusers (rod diffuser and output diffuser), and a reflective cavity, quality Gbps visible light communication was achieved at distances up to 3.81 meters and receiver angles as large as 30 degrees. These results demonstrate that it is possible to have high-speed VLC with efficient, speckle-free (<5%) white light from multi-stage scattering and reflection, achieving bit error rates as low as 2.4×10^{-11} at a distance of 1 meter.

Acknowledgements. This work was supported by the National Science Foundation (CNS- 2308686) and the National Science Foundation Research Traineeship, Transformative Research and Graduate Education in Sensor Science, Technology and Innovation (DGE- 2125733). Any opinions, findings, and conclusions or recommendations expressed in this material are those of the author(s) and do not necessarily reflect the views of the National Science Foundation.

Disclosure of Interests. The authors have no competing interests to declare that are relevant to the content of this article.

References

1. International Energy Agency, https://www.iea.org/energy-system/buildings/lighting#tracking, last accessed 2025/01/22.
2. Chenxin Liang, Jiarong Li, Sicong Liu et al. Integrated sensing, lighting and communication based on visible light communication: A review, 2024, Digital Signal Processing Volume 145 [https://doi.org/10.1016/j.dsp.2023.104340.]
3. Jifeng Liu, Xiaoxin Wang, Charles Carver et al. High-efficiency, low-speckle contrast white laser lighting via multi-stage scattering and photon recycling, 03 August 2023, PREPRINT (Version 1) available at Research Square [https://doi.org/10.21203/rs.3.rs-3182555/v1]
4. M. Pattison, M. Hansen, N. Bardsley et al. U.S. Department of Energy, 2022 Solid-State Lighting R&D Opportunities, February 2022. https://www.energy.gov/eere/ssl/articles/2022-solid-state-lighting-rd-opportunities
5. Advanced Micro Devices, https://www.xilinx.com/products/intellectual-property/ibert_ultrascale_gth.html, last accessed 2025/01/22.
6. Junhui Hu, Fangchen Hu, Junlian Jia et al. 46.4 Gbps visible light communication system utilizing a compact tricolor laser transmitter, 2022, Opt. Express 30, 4365-4373

Evaluation of Module for Alignment-Free Diffuser Film Composed of Dielectric Multilayers and Microprisms for Mini LED Backlight

Marii Nishikawa [1,2], Yukio Taniguchi [1], Masahiro Goto [1], Yasuyuki Ohyagi [1], Yoshihiro Kanai [1], Masayuki Sekido [1], Hirotsugu Yamamoto [2]

[1] Dai Nippon Printing Co., Ltd., 250-1 Wakashiba, Kashiwa-shi, Chiba 277-0871, Japan
[2] Utsunomiya University, 7-1-2 Yoto, Utsunomiya-shi, Tochigi 321-0904, Japan
Nishikawa-M8@mail.dnp.co.jp

Abstract. Three types of diffuser films for mini-LED backlight are developed and their optical performances are examined. The proposed diffuser films are composed of a dielectric multilayer and microprisms and features alignment free. This paper explains a module evaluation using the diffuser film designed for mini-LED backlights. The diffuser film changes the light emitted from the LED and the light reflected by the dielectric multilayer. The dielectric multilayer function as an angle filter that does not transmit blue light directly above the LED and transmits blue light only with an angle of incidence above a certain degree. The microprisms divide and refract light emitted from LED light and multiplies the LED light laterally. This combination of dielectric multilayer and microprisms makes the illuminance distribution uniform without requiring alignment between the LED chips and the diffuser film. In this paper, we discuss the film constitution to further enhance the effects of the diffuser film. In particular, we will explain the module evaluation results when a microprisms are combined with the light-emitting side of the dielectric multilayer, as well as when it is laminated with a wavelength conversion film.

Keywords: mini-LED backlight, diffuser film, alignment free.

1 Introduction

The Mini-LED backlight, characterized by its thin module, high luminance, high contrast, and low power consumption, is anticipated to be the next generation display method [1]. The diffuser film is one of the key components for mini-LED backlight system. Microstructures in the diffuser film can be make the backlight module thin [2]. Challenges on the diffuser films are to reduce the mura of LED and make the backlight uniform [3]. Recently, we have proposed a diffuser film that is composed of dielectric multilayers and microprisms [4, 5]. This composite structure makes the mini-LED backlight uniform without precise alignment. The dielectric multilayers contribute for selective blue light transmission. The microprisms layer contributes for LED light

S. Kurimura and T. Yagi (eds.), *Proceedings of Laser Display and Lighting Conference 2025*, Springer Proceedings in Physics 429,
https://doi.org/10.1007/978-981-96-9782-3_35

direction control. This combination ensures uniform illuminance without the necessity for LED-diffuser alignment and improves efficiency.

In this paper, module evaluation using the diffuser film for mini-LED backlight is proposed. The basic configuration of the diffuser film and backlight used for the evaluation is shown in Fig. 1. There is possibility for use of double microprism layers to improve uniformity. Furthermore, optimization of sandwich structures may improve light utilization efficiency. The purpose of this paper is to compare multiple constitution performance of mini-LED backlight using the diffuser film. Since the light emitted from the diffuser film has a large emission angle due to the characteristics of the angle filter, the emission angle is refracted in the low direction to further improve luminance and uniformity. In particular, the effect of adding microprisms and eliminating the air interface between the QD film and the QD film is investigated.

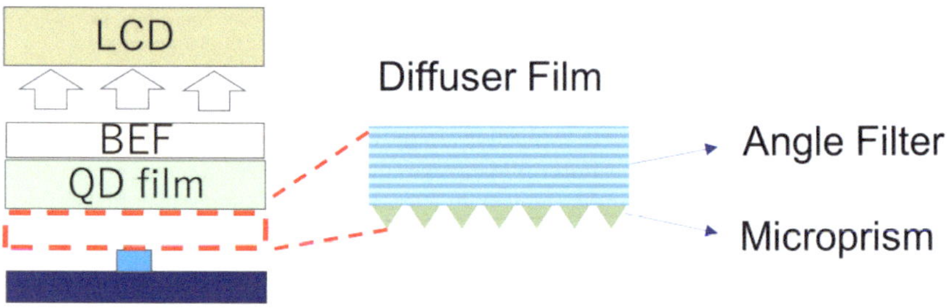

Fig. 1. Composition of mini-LED backlight and diffuser film

2 Principle of Diffuser Film

The proposed diffuser film is composed of a multilayer dielectric and a microprism layer. The dielectric layers function as an angle filter that does not transmit blue light directly above the LED and only transmits blue light with an incident angle above a designed value. The microprism layer controls the direction of the LED light and replicates the LED lights laterally. This combination of angle filter and microprism layer makes the illuminance distribution uniform without requiring alignment between the LED chips and the diffuser film. Our proposed diffuser film provides 4x better efficiency, better uniformity than conventional films, and only 50 μm thickness [6].

The configurations examined in this study are shown in Fig. 2(a), (b), (c), and (d). The microprisms pattern is shown in Fig. 3 (a) and (b). Pattern 1 is provided on the LED side of all diffuser films, and Pattern 2 is provided on the QD film side of Type 2.

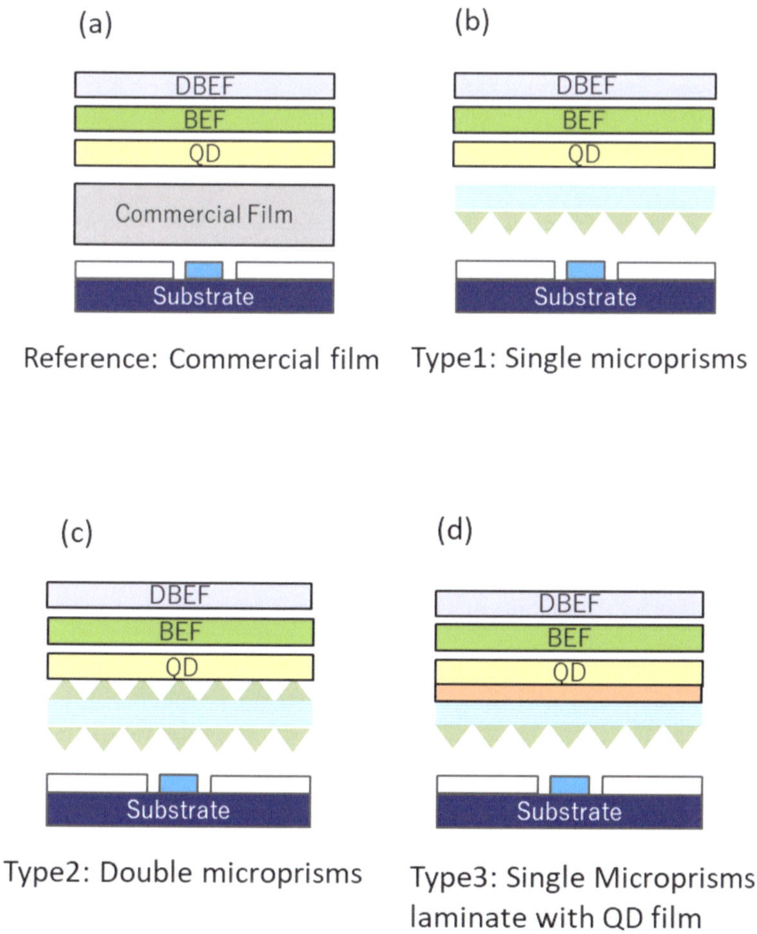

Fig. 2. Configurations of diffuser films examined in this study

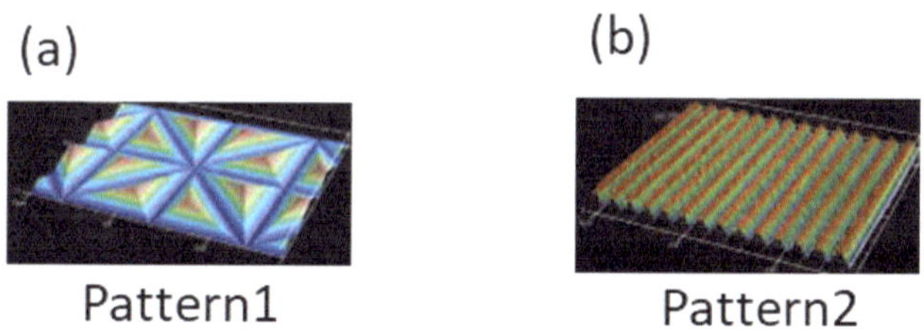

Fig. 3. Microprisms patterns

3 Experiments

For the diffuser films, described in the previous section, were fabricated and used with an actual mini-LED backlight module for performance evaluation experiments. The LED size is 0.1 mm × 0.2 mm×0.085 mm(height). The LED pitch is 4 mm. In the part without LEDs, a diffusive reflector was implemented.

We have evaluated a reference and three types of diffuser films. A commercial film was used for a reference and mounted at a position of 0.5 mm from the LED top. In Type 1, single microprisms were mounted. In Type 2, double microprisms were mounted. In Type 3, QD film was laminated to single microprisms.

We have measured luminance distributions of the four types of mini-LED backlight systems. A 2D spectroradiometer (The SR5000 manufactured by Topcon Technohouse) was used for module evaluation. The results of luminance distributions are shown in Fig. 4. Compared to the commercial film, it can be seen that the diffuser films with microprisms give uniform luminance distributions. Use of double microprisms gives more uniform distribution. Light utilization efficiency and Uniformity are shown in Fig. 5. Lamination with the QD layer with the microprisms results the highest light utilization efficiency. Uniformity of Type 2 and Type 3 are almost as the same.

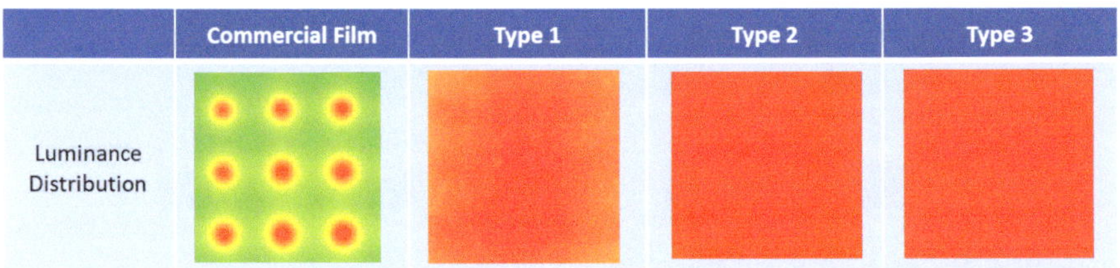

Fig. 4. Luminance distributions

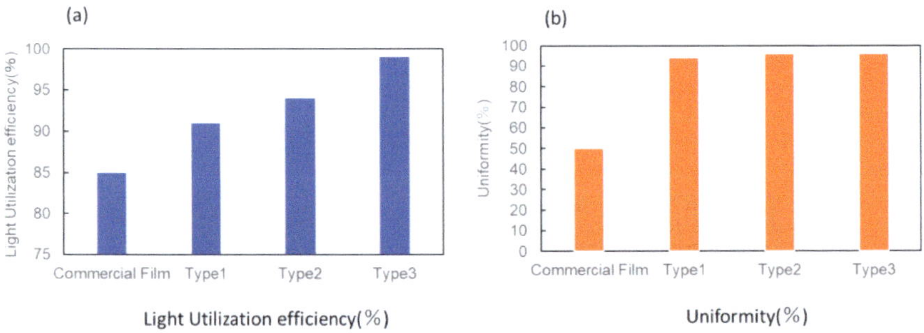

Fig. 5. Result of (a) light utilization efficiency and (b) uniformity

Thus, experiments show that we were able to realize a backlight module that is thin and has high light utilization efficiency and uniformity without the need for alignment.By using the double microprisms type, it is possible to further improve the light utilization efficiency and uniformity.By laminating the single microprisms type with the QD film on the light emitting side, it is possible to improve the light utilization efficiency and uniformity.

4 Conclusion

We have introduced a module evaluation of multiple constitution using the diffuser film composed of a dielectric multilayer angle filter and a microprisms. Comparing 3 types of component structures, by using double microprisms or laminating with QD film, measured the backlight module efficiency as 94~99% and uniformity as 96%. Thus, we have succeeded in realizing a thin, efficient, alignment-free diffuser film. Alignment-free assembly allows higher productivity manufacturing during backlight assembly. By using this diffuser film, it is possible to obtain a mini-LED backlight display with high contrast, high brightness, and low power consumption, which enables local dimming control with an unprecedented thinness.

References

1. Xiao J, Fei J, Zheng F, Liu Q, Huo W, Li J, Mei W, Xu H, Zhang S. Mini-LED backlight units on glass for 75-inch 8K resolution liquid crystal display. Journal of the Society for Information Display, 30, 54–60 (2022).
2. Feng Q, Xiao H, Yang L, Zhu B, Yuan Y, Wang Z, Lv G. Ultrathin miniLED backlight system using optical film with microstructures. Applied Optics, 60(30), 9497–9503 (2021).
3. Hu Y, Hu L, Yuan L, Chen E, Xu S, Guo T, Ye Y. Design of microstructure optical film for high uniform mini-LED backlight. Proc SPIE, 12765, 127650B (2023).
4. Nishikawa M, Kanai Y, Sekido M. Thin diffuser film that does not require alignment for local dimming mini-LED display. Proc. IDW, 30, 492–494 (2023).
5. Nishikawa M, Taniguchi Y, Goto M, Ohyagi Y,Kanai Y, Sekido M, Yamamoto H, SPIE Photonics West 2025 ,Paper 13389-23 (2025).
6. Nishikawa M, Taniguchi Y, Goto M, Ohyagi Y,Kanai Y, Sekido M, Yamamoto H, Journal of the Society for Information Display, in press. https://doi.org/10.1002/jsid.2021

Elimination of Hot Pixels in Measuring the Point Spread Function of Aerial Display

Takumi Watanabe[1], Kazuaki Takiyama[1], Shiro Suyama[1], and Hirotsugu Yamamoto[1]

[1] Utsunomiya University, 7-1-2 Yoto, Utsunomiya, Tochigi, Japan
hirotsugu@yamamotolab.science

Abstract. Aerial images are formed with anisotropy due to diffraction from the array of micro-optics that compose the aerial imaging optics. The point spread function (PSF) is used as an indicator to evaluate the spatial resolution of aerial imaging optics, and 4 times standard deviation (D4σ), which can be adapted to non-Gaussinan distributions, is considered appropriate to represent the aerial image diameter. However, abnormal values called hot pixels, which do not perform their normal function on the imaging sensor have a significant impact on the output value of D4σ in terms of calculation. In this study, we aimed to clarify the usefulness of hot pixels elimination. To achieve this, hot pixels are defined as hot pixels as pixels that are above the top 0.01% of the maximum luminance value and compare the D4σ results before and after elimination. As a result, the accuracy of D4σ is confirmed to be improved by eliminating hot pixels. Therefore, the elimination of hot pixels is considered to be an effective method to improve the accuracy of PSF quantification.

Keywords: Aerial display, Hot pixels, Point spread function.

1 Introduction

Aerial displays form a real image in mid-air and are expected to be applied as novel non-contact interfaces. The aerial images are formed using real-image-forming optical system, which is composed of an array of micro-optics. However, the image quality of the aerial image in specific directions is sometimes degraded by diffraction caused by the array of micro-optics. Additionally, there is no established international standard for evaluating the quality of aerial images.

Aerial imaging by retro-reflection (AIRR) [1] and dihedral corner reflector array (DCRA) [2] have been widely used as major methods to form aerial images. The Point spread function (PSF), which is a performance indicator of display optics, is expected to be used to evaluate the image quality of aerial images. PSF can spatial distribution, including diffractions in certain directions and is used for the international standard on measurement method of light-emitting diode (LED) backlight unit [3]. Recently, a measurement method has been proposed to quantify the spatial resolution of augmented reality head-mounted displays by separating the effects of optics, display, and rendering pipeline [4]. This measurement method of capturing and fitting PSFs of various sizes can also be applied to the measurement of PSFs of aerial imaging optics. In a previous study,

S. Kurimura and T. Yagi (eds.), *Proceedings of Laser Display and Lighting Conference 2025*, Springer Proceedings in Physics 429,
https://doi.org/10.1007/978-981-96-9782-3_36

based on the method, we clarified that D4σ has been shown to be suitable for showing the PSF of an aerial display optical system [5]. However, the D4σ results were not accurate due to noise effects such as hot pixels outputting abnormal values. Therefore, the elimination of hot pixels is necessary for a more quantitative PSF measurement.

The objective of this study is to clarify the usefulness of eliminating hot pixels. We define hot pixels and show their usefulness with D4σ results after elimination.

2 Principles of Aerial Display Optical Systems

Two types of aerial imaging techniques used in this study are shown in Fig. 1. The principle of AIRR is shown in Fig. 1(a). AIRR consists of a light source, a beam splitter, and a retro-reflector. The light emitted from the light source is reflected by the beam splitter and impinges on the retro-reflector. The light incident is retro-reflected by the retro-reflector and returned to the beam splitter. The light passing through the beam splitter forms an aerial image at a plane-symmetrical position of the light source.

The principle of DCRA is shown in Fig. 1(b). Light emitted from a light source is reflected twice by two adjacent vertical mirror surfaces in each element in the DCRA. The light reflected twice forms the aerial image at the plane-symmetrical position of the light source. DCRA consists of only two optical components and has a short overall optical path length.

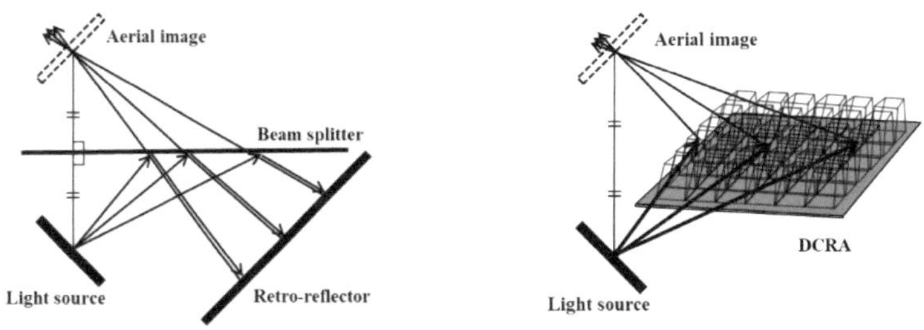

(a) Principle of AIRR. (b) Principle of DCRA.

Fig. 1. Principles of two types of aerial display optical systems.

3 Hot Pixels Detection

A hot pixel is a pixel that does not function properly and outputs an abnormal value due to localized defects. We estimated the hot pixels in a light measuring device (Radiant Vision Systems: Prometric Y-29) used in this study from sample luminance values data. The sample data was obtained by closing the cap of the lens attached to the LMD, setting the exposure time to automatic, and measuring 30 times. The number of sample data corresponds to the number of pixels in the LMD, so there are 6576*4384 sample data. The luminance values of the central horizontal and central vertical axes of the sample data are shown in Fig. 2. In Fig. 2, most pixels have a luminance value close t0 cd/m^2. However, there are some pixels with clearly high luminance values. A histogram of the sample data is shown in Fig. 3. The vertical axis is expressed in logarithms. In

Fig. 3, Pixels can be seen scattered on the side of high luminance values from 1 cd/m^2 to 5 cd/m^2. In this study, we defined hot pixels as pixels that are above the top 0.01% of the maximum luminance value in the sample data.

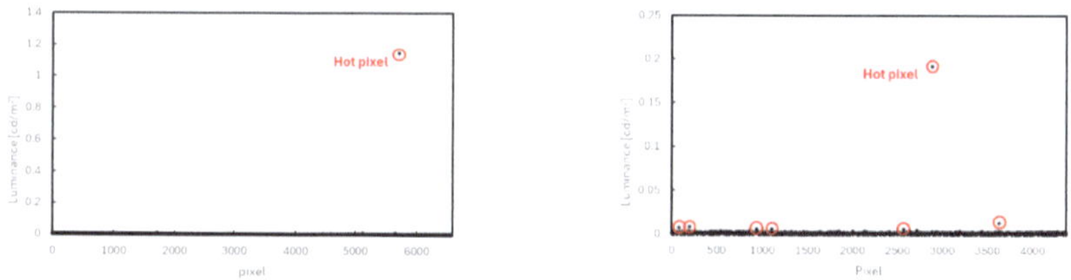

(a) Central horizontal axis. (b) Central vertical axis.

Fig. 2. Luminance values of sample data.

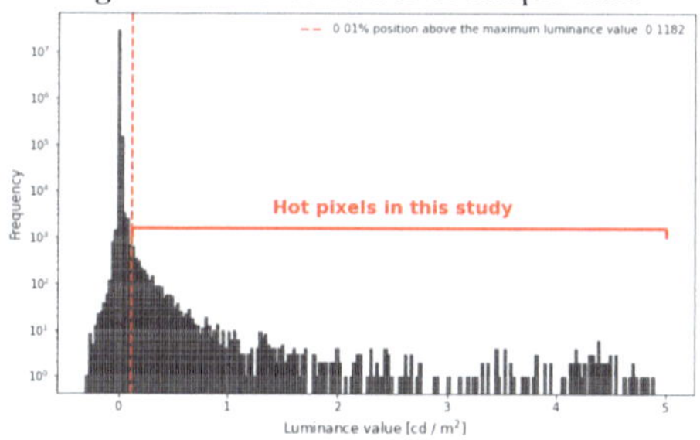

Fig. 3. Histogram of the sample luminance values data.

4 D4σ Measurement of PSF in Plane-Symmetric Real-Image-Forming Optics for Aerial Display

4.1 Experimental Setup and Method

The PSF of AIRR and DCRA optics was measured using circular light sources in different sizes. D4σ is derived from the 2D luminance distribution of an aerial image. Apply fitting to each value for aperture size to obtain the intercept value. The derived intercept values show the PSF of only the array of micro-optics, as the light source size is 0.

The experimental setups for AIRR and DCRA optical system are shown in Fig. 4. An integrating sphere with an aperture stop was used as the light source. Aerial images were taken by changing the aperture size to 1mm, 2mm, 3mm, 4mm, 5mm, and 6 mm with a floating distance of 100mm. The LMD was used to capture aerial images. An imaging lens which a focal length of 100 mm was attached to the LMD. The focal length, f-number, exposure time, and number of measurements of the LMD camera were set to 325 mm, 8.0, automatic, and 10 times.

In the AIRR optical system, a half-mirror was used as a beam splitter, and a prism-type retro-reflector (Nippon Carbide Industries: RF-Ax) was used as a retro-reflector. The distance from the beam splitter to the retro-reflector was 50 mm. Note that the direction of the retro-reflector was tilted by 5 degrees to avoid overlapping a ghost image that is caused by reflection on the retro-reflector surface.

In the DCRA optical system, a DCRA (Parity Innovations: a 300-mm test sample with a light-shielding mask) was used to form an aerial image. The aerial images for each aperture of the AIRR and DCRA are shown in Fig. 5, 6. A comparison of the aerial images shows that the diffraction direction is different.

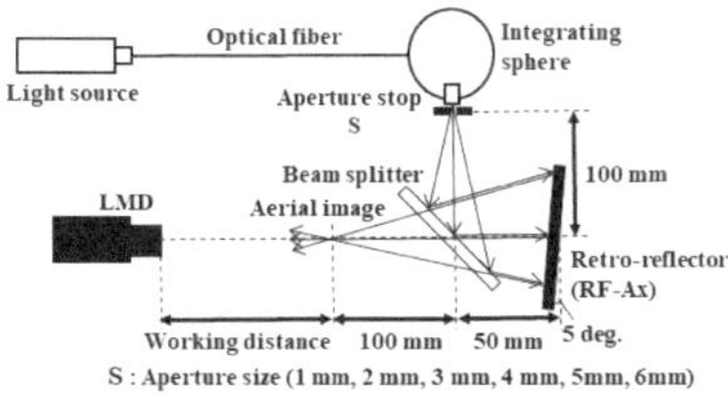

(a) AIRR optical system

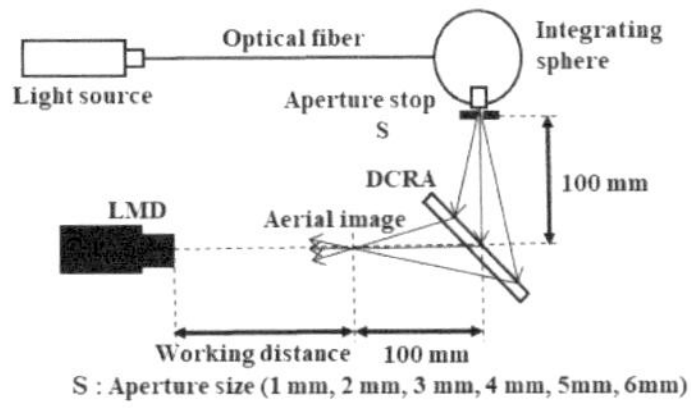

(b) DCRA optical system

Fig. 4. Experimental setup of aerial display.

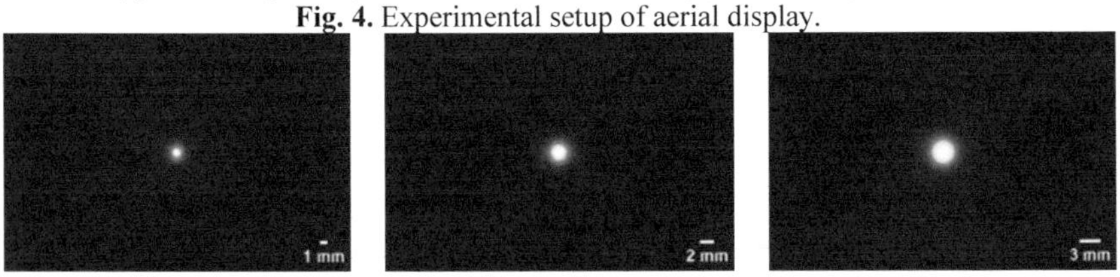

(a) S = 1 mm. (b) S = 2 mm. (c) S = 3 mm.

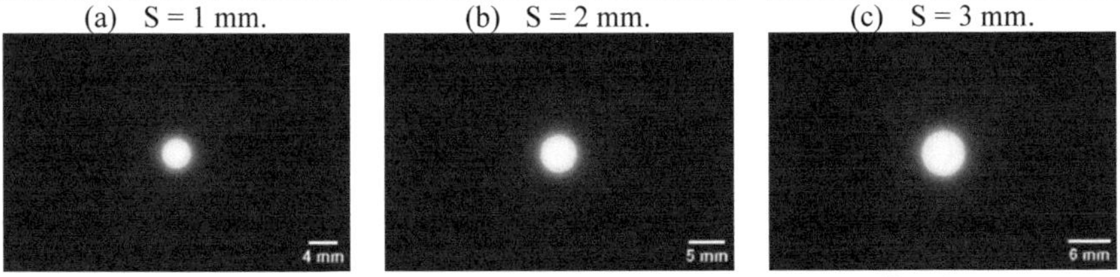

(d) 4 mm. (e) 5 mm. (f) 6 mm.

Fig. 5. Aerial images of the AIRR optical system.

(a) S = 1 mm. (b) S = 2 mm. (c) S = 3 mm.

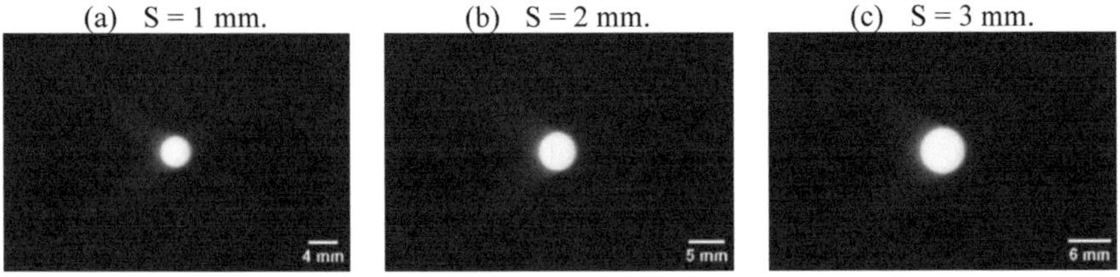

(d) 4 mm. (e) 5 mm. (f) 6 mm.

Fig. 6. Aerial images of the DCRA optical system.

4.2 D4σ Results for Hot Pixels elimination

The D4σ of PSF is derived as follows. The average value $\bar{x}$ of the horizontal component x of the data can be calculated as follows:

$$\bar{x} = \frac{\int_{-\infty}^{\infty}\int_{-\infty}^{\infty} I\,(x, y)x\;dxdy}{\int_{-\infty}^{\infty}\int_{-\infty}^{\infty} I\,(x, y)\;dxdy}, \tag{1}$$

Where $I(x, y)$ denotes the luminance value of x and y components, Then, D4σ is expressed by

$$D4\sigma = 4\sqrt{\frac{\int_{-\infty}^{\infty}\int_{-\infty}^{\infty} I(x, y)\;(x - \bar{x})^2\;dxdy}{\int_{-\infty}^{\infty}\int_{-\infty}^{\infty} I(x, y)\;dxdy}}. \tag{2}$$

This formula can be applied to the vertical component as well. In this paper, less than $1/e^2$ (13.53%) of the maximum luminance value was eliminated as background noise. Additionally, the defined hot pixels are also eliminated. The elimination does not include the D4σ calculation process.

D4σ results of the AIRR and DCRA optical system with the aperture sizes are shown in Fig. 7. The result in Fig. 7(a) is obtained when hot pixels are not eliminated in the calculation of D4σ. The intercept values when fitting each D4σ value are 1.7974 mm (horizontal axis) and 1.2819 mm (vertical axis) for AIRR, and 0.8903 mm (horizontal axis) and 0.4348 mm (vertical axis) for DCRA. The result in Fig. 7(b) is obtained when hot pixels are eliminated in the calculation of D4σ. The intercept values when fitting each D4σ value are 0.5138 mm (horizontal axis) and 0.7336 mm (vertical axis) for AIRR, and 0.467 mm (horizontal axis) and 0.2312 mm (vertical axis) for DCRA. The fitting in Fig. 7(b), where the hot pixels are not included in the calculation of D4σ, is linear and has smaller intercept values.

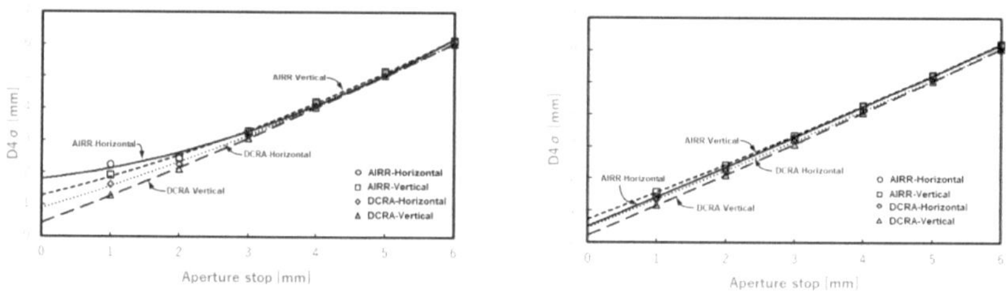

(a) Before hot pixels elimination. (b) After hot pixels elimination.

Fig. 7. D4σ results of the AIRR and DCRA optical system with the aperture sizes.

5 Discussion

We discuss the improvement in D4σ accuracy due to hot pixels elimination. In Fig. 7(a), the result before hot pixels elimination, the D4σ value fitting graphs at each aperture size are curvilinear and the intercept values, which should represent only diffracted light, are very large. On the other hand, in Fig. 7(b), the result after hot pixels elimination, the D4σ value fitting graphs are linear and the intercept values are small. This is because D4σ is greatly affected by the luminance values at the edges of the distribution, and the accuracy of the calculation is thought to have been improved hot pixels elimination.

6 Conclusion

We defined hot pixels and compared D4σ results before and after elimination. As a result, the D4σ value fitting graphs for each aperture size of the AIRR and DCRA optics are linear and the intercept values are small. Therefore, hot pixels elimination is considered to be an effective method of quantifying PSF measurements.

Acknowledgements

A part of this work was supported by Japan Society for the Promotion of Science (20H05702, 24KJ0518) and JST (grant no. JPMJMI24H1).

References

1. Yamamoto, H., Tomiyama, Y., and Suyama, S.: Floating aerial LED signage based on aerial imaging by retro-reflection (AIRR). Opt. Express 22, pp. 26919–26924 (2014).
2. Maekawa, S., Nitta, K., and Matoba, O.: Transmissive optical imaging device with micromirror array. Proc. SPIE. Vol. 6392, pp. 130–137 (2006).
3. International Electrotechnical Commission.: Display lighting unit – Part 2-1: Electro-optical measuring methods of LED backlight unit. International Standard. IEC 62595-2-1:2016 (2016).
4. Johnson,M., Zhao, C., Varshney, A., and Beams, R.: Quantifying the optical and rendering pipeline contributions to spatial resolution in augmented reality displays. J.SID, 32 (8), pp. 555–567 (2024).
5. Watanabe, T., Takiyama, K., Iwane, T., Suyama, S., Nishikawa, M., Kikuta, H., Naka, K., Uehara, S., and Yamamoto, H.: Point Spread Measurement of Plane-Symmetric Real-Image-Forming Optics for Aerial Display. Proc. IDW'24. Vol. 31, pp. 441–444 (2024).

MIX
Papier aus verantwortungsvollen Quellen
Paper from responsible sources
FSC® C105338

If you have any concerns about our products,
you can contact us on
ProductSafety@springernature.com

In case Publisher is established outside the EU,
the EU authorized representative is:
Springer Nature Customer Service Center GmbH
Europaplatz 3, 69115 Heidelberg, Germany

Printed by Libri Plureos GmbH
in Hamburg, Germany